3rd University Scholar Conference (USC 2023)

IOP Conference Series: Materials Science and Engineering
Volume 1314

Dhulikhel, Nepal
4-5 October 2023

ISBN: 979-8-3313-0661-8
ISSN: 1757-8981

Printed from e-media with permission by:

Curran Associates, Inc.
57 Morehouse Lane
Red Hook, NY 12571

Institute of Physics
Dirac House, Temple Back
Bristol BS1 6BE UK

Phone: 44 1 17 929 7481
Fax: 44 1 17 920 0979

techtracking@iop.org

Additional copies of this publication are available from:

Curran Associates, Inc.
57 Morehouse Lane
Red Hook, NY 12571 USA
Phone: 845-758-0400
Fax: 845-758-2633
Email: curran@proceedings.com
Web: www.proceedings.com

TABLE OF CONTENTS

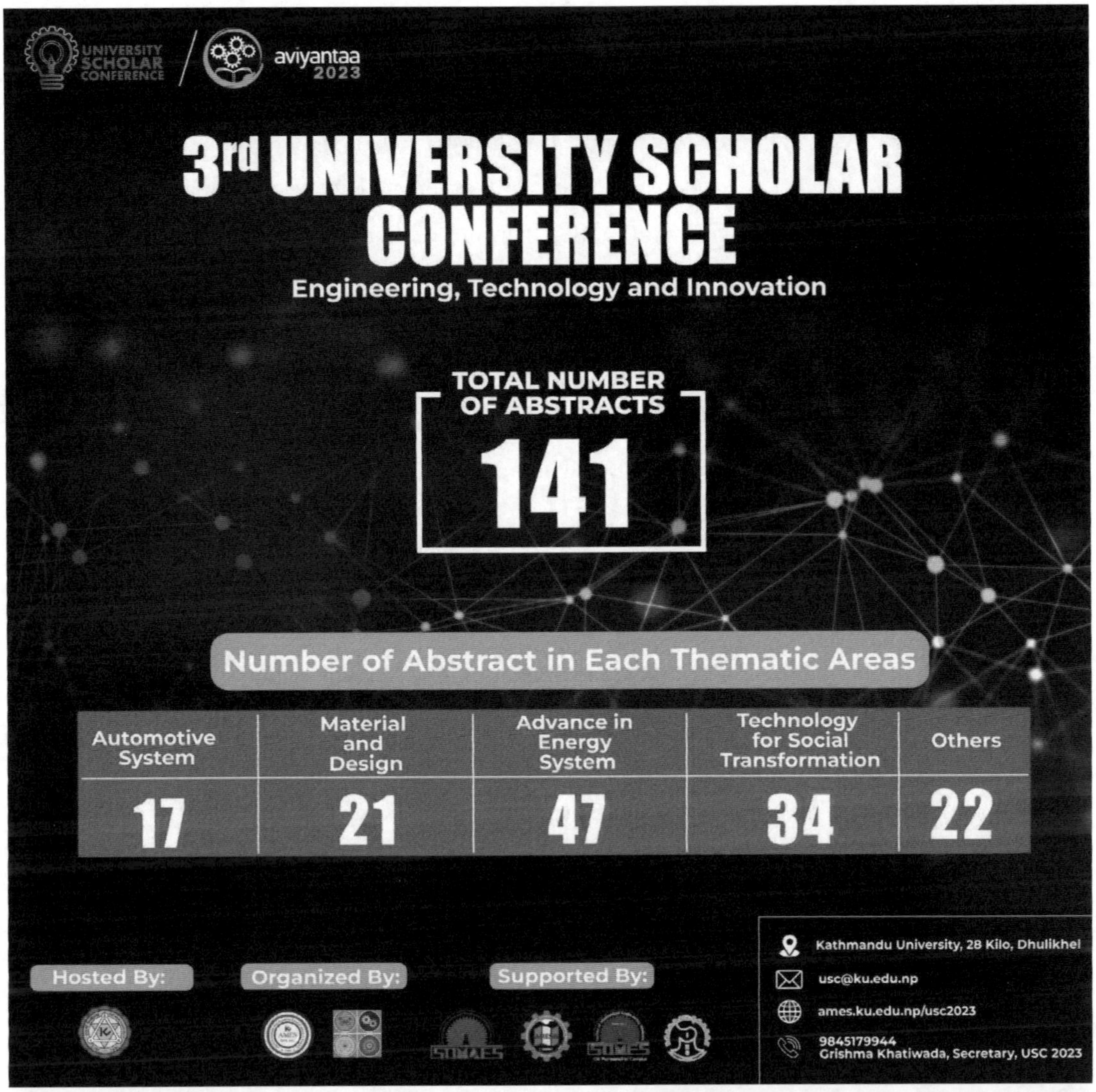
UNIVERSITY SCHOLAR CONFERENCE
aviyantaa 2023
3rd UNIVERSITY SCHOLAR CONFERENCE
Engineering, Technology and Innovation
TOTAL NUMBER OF ABSTRACTS
141
Number of Abstract in Each Thematic Areas
Automotive System
17
Material and Design
21
Advance in Energy System
47
Technology for Social Transformation
34
Others
22
Hosted By:
Organized By:
Supported By:
Kathmandu University, 28 Kilo, Dhulikhel
usc@ku.edu.np
ames.ku.edu.np/usc2023
9845179944
Grishma Khatiwada, Secretary, USC 2023

KATHMANDU UNIVERSITY, DHULIKHEL, NEPAL Tel No: 011-415100 ku.edu.np

IOP Conf. Series: Materials Science and Engineering **1314** (2024) 011001 doi:10.1088/1757-899X/1314/1/011001

Preface

The University Scholar Conference: Engineering and Innovation (USC) is a conference organized by the Association of Mechanical Engineering (AMES) at Kathmandu University (KU). The first edition, USC 21, was successfully completed in virtual mode, the second edition, USC 22, was held in a hybrid mode, and third edition, USC 2023, was held on 4-5 October 2023. USC 2023 was hosted by Kathmandu University, organized by AMES and supported by the Department of Mechanical Engineering – KU and the University Grant Commission (UGC). The inauguration ceremony of the two-day conference event was held at Kathmandu University with a keynote session delivered by Prof. Sujatha Srinivasan, IIT Madras, and technical remarks from Prof. Sachin L. Borse and Prof. Catherine Holloway.

The University Scholar Conference 2023 provided an excellent platform for graduate and undergraduate students to share their cutting-edge developments and state-of-the-art research work in the fields of science and engineering. One interesting fact about this conference is that it was completely organized by a student club under the supervision of faculty. Thus, the USC conference is a student-level conference *organized by university students for university students*. The conference was attended by a mix of academia, researchers, industrialists, professors, students, and policymakers. The conference committee highly encouraged the participation of graduate and undergraduate students engaged in engineering and science education within Nepal and abroad.

With the tagline "*Engineering, Technology and Innovation*" four thematic areas were defined for USC 22: i) Automotive Systems, ii) Materials and Design for Transformation, iii) Advances in Energy Systems, and iv) Technology for Social Transformation. A total of 141 abstracts were received at the initial phase, out of which 48 papers were selected for presentation in the conference. Participants from different universities and different countries presented their research works in 12 plenary sessions during the two conference days. In addition to the parallel plenary sessions, a poster presentation was also conducted for funded projects. Based on the quality of the presented work, four best paper awards were declared for each thematic area, and a gift hamper was presented to the winners. The 48 papers presented at the conference were considered for further publication in the *IOP Conference Series: Materials Science and Engineering*. After a detailed peer-review process, a total of 15 papers were selected for final publication in this volume.

IOP Conf. Series: Materials Science and Engineering **1314** (2024) 011001 doi:10.1088/1757-899X/1314/1/011001

Overall, the 3rd University Scholar Conference: Engineering, Technology and Innovation (USC 2023) was a success and provided a valuable platform for university scholars to share their research and engage with other professionals in the field of science and engineering. We hope that this conference will continue to be organized and provide opportunities for students to showcase their work and enhance their academic careers. We would like to express our sincere gratitude to the members of the Scientific Committee and Advisory Committee who made great efforts to review and revise the conference papers, ensuring the quality and scientific level of the papers. We hope that the proceedings will promote similar research practices at the undergraduate level.

Surendra Sujakhu, PhD
Editor-in-Chief and Coordinator, University Scholar Conference 2023
Faculty Coordinator, Association of Mechanical Engineering Students, Nepal
Assistant Professor, Department of Mechanical Engineering, Kathmandu University, Nepal

Ranjit Shrestha
Editor, University Scholar Conference 2023
Assistant Professor, Department of Mechanical Engineering, Kathmandu University, Nepal

IOP Conf. Series: Materials Science and Engineering **1314** (2024) 011001 doi:10.1088/1757-899X/1314/1/011001

Conference Chair

Dr. Hari Prasad Neopane	Professor, Department of Mechancial Engineering, Kathmandu University, Dhulikhel, Nepal

Organizing Committee

Dr. Surendra Sujakhu	Conference Coordinator	Kathmandu University, Nepal
Mr. Grishma Khatiwada	Conference Secretary	Kathmandu University, Nepal
Mr. Utsav Pokharel	Technical Management Chair	Kathmandu University, Nepal
Mr. Madan Ghimire	Technical Management Chair	Kathmandu University, Nepal
Mr. Sugam Karki	Technical Management Chair	Kathmandu University, Nepal
Mr. Sangam Bhandari	Publicity Chair	Kathmandu University, Nepal
Ms. Yarana Rai	Publicity Chair	Kathmandu University, Nepal
Mr. Yogesh Dumre	Publication Chair	Kathmandu University, Nepal
Mr. Sarams Siwakoti	Publication Chair	Kathmandu University, Nepal
Mr. Prasanna Kshetri	Registration Chair	Kathmandu University, Nepal
Mr. Sandesh Paudel Chettri	Registration Chair	Kathmandu University, Nepal
Mr. Anmol Shrestha	Web Service Chair	Kathmandu University, Nepal
Mr. Akhilesh Kafle	Web Service Chair	Kathmandu University, Nepal
Mr. Prabin Bhattrai	Corporate and Finance Chair	Kathmandu University, Nepal
Ms. Sadikxya Pandey	Corporate and Finance Chair	Kathmandu University, Nepal
Mr. Milish Dhungel	President, AMES	Kathmandu University, Nepal
Mr. Manish R.C.	Secretary, AMES	Kathmandu University, Nepal
Mr. Krishal Dhamala	Executive, AMES	Kathmandu University, Nepal

Advisory Committee

Dr. Bhola Thapa	Vice-chancellor, Professor, Kathmandu University, Nepal
Dr. Bim Prasad Shrestha	Professor, Kathmandu University, Nepal
Dr. Bivek Baral	Professor, Kathmandu University, Nepal
Dr. Manish Pokharel	Professor, Kathmandu University, Nepal
Dr. Janardhan Lamichane,	Professor, Kathmandu University, Nepal
Dr. Bhupendra Bimal Chhetri	Professor, Kathmandu University, Nepal
Dr. Daniel Tuladhar	Professor, Kathmandu University, Nepal
Dr. Rajendra Joshi	Professor, Kathmandu University, Nepal
Dr. Sunil Prasad Lohani	Associate Professor, Kathmandu University, Nepal
Dr. Krishna P. Shrestha	Associate Professor, Kathmandu University, Nepal

Dr. Biraj Singh Thapa	Associate Professor, Kathmandu University, Nepal
Dr. Binay K.C.	Associate Professor, Kathmandu University, Nepal
Dr. Shyam Sundar Khadka	Associate Professor, Kathmandu University, Nepal
Dr. Subash Ghimire	Associate Professor, Kathmandu University, Nepal
Dr. Bal Krishna Bal	Associate Professor, Kathmandu University, Nepal
Dr. Shailendra Kumar Jha	Assistant Professor, Kathmandu University, Nepal
Dr. Khagendra Acharya	Assistant Professor, Kathmandu University, Nepal
Dr. Sailesh Chitrakar	Assistant Professor, Kathmandu University, Nepal
Dr. Ranjit Shrestha	Assistant Professor, Kathmandu University, Nepal
Mr. Pratisthit Lal Shrestha	Assistant Professor, Kathmandu University, Nepal
Mr. Malesh Shah	Assistant Professor, Kathmandu University, Nepal
Mr. Bijendra Shrestha	Assistant Professor, Kathmandu University, Nepal
Mr. Hari Dhakal	Assistant Professor, Kathmandu University, Nepal
Mr. Niranjan Bastakoti	Assistant Professor, Kathmandu University, Nepal
Mr. Chiranjeevi Mahat	Assistant Professor, Kathmandu University, Nepal
Mr. Gokarna Paudel	Assistant Professor, Kathmandu University, Nepal
Mr. Pawan Karki	Assistant Professor, Kathmandu University, Nepal
Mr. Jahid Ahmed Jibran	Assistant Professor, Kathmandu University, Nepal
Mr. Uren Amatya	Lecturer, Kathmandu University, Nepal
Ms. Sirapa Shrestha	Teaching Assistant, Kathmandu University, Nepal
Mr. Sagar Pandit	Teaching Assistant, Kathmandu University, Nepal

Scientific Committee

Dr. Jeeban Poudel	University of Queensland, Australia
Dr. Milan Shrestha	Nanyang Technological University, Singapore
Dr. Anil Kumar Bastola	University of Nottingham, United Kingdom
Dr. Niroj Maharjan	Advance Remanufacturing and Technology Center, Singapore
Dr. Milan Paudel	Nanyang Technological University, Singapore
Dr. Arun Satyal	Nepal Engineering College, Nepal
Mr. Deepak Paudel	Aalto University, Finland
Mr. Janak Aryal	Wärtsilä Finland Oy
Dr. Bikash Adhikari	Kathmandu University, Nepal
Dr. Bishal Silwal	Kathmandu University, Nepal
Dr. Nawaraj Sanjel	Kathmandu University, Nepal
Dr. Sushil Shrestha	Kathmandu University, Nepal
Dr. Binay KC	Kathmandu University, Nepal
Dr. Krishna P. Shrestha	Kathmandu University, Nepal

IOP Conf. Series: Materials Science and Engineering **1314** (2024) 011001 doi:10.1088/1757-899X/1314/1/011001

Dr. Biraj Singh Thapa	Kathmandu University, Nepal
Mr. Biraj Dhungana	Green Road Waste Management, Nepal
Mr. Abhishek Karki	Abhiyantrika Karmashala, GDS Fiber World, Nepal
Dr. Parash Mani Timilsina	Kathmandu University, Nepal
Dr. Surendra Sujakhu	Kathmandu University, Nepal
Dr. Ranjit Shrestha	Kathmandu University, Nepal
Dr. Sailesh Chitrakar	Kathmandu University, Nepal
Mr. Pratisthit Lal Shrestha	Kathmandu University, Nepal
Mr. Malesh Shah	Kathmandu University, Nepal
Mr. Bijendra Shrestha	Kathmandu University, Nepal
Mr. Hari Dhakal	Kathmandu University, Nepal
Mr. Chiranjeevi Mahat	Kathmandu University, Nepal
Mr. Pawan Karki	Kathmandu University, Nepal
Mr. Gokarna Paudel	Kathmandu University, Nepal
Mr. Jahir Ahmed Jibran	Kathmandu University, Nepal
Mr. Nischal Chaulagain	Kathmandu University, Nepal
Mr. Uren Amatya	Kathmandu University, Nepal
Ms. Sirapa Shrestha	Kathmandu University, Nepal
Mr. Sagar Pandit	Kathmandu University, Nepal

Conference Session Chairs

Dr. Milan Poudel	Nanyang Technological University, Singapore
Dr. Anil Kumar Bastola	University of Nottingham, United Kingdom
Dr. Prajib Joshi	Western Regional Campus, Nepal
Dr. Bijay Thapa	EnergizeNepal, Nepal
Dr. Ranjit Shrestha	Kathmandu University, Nepal
Dr. Binaya K.C.	Kathmandu University, Nepal
Dr. Bishal Silwal	Kathmandu University, Nepal
Dr. Sailesh Chitrakar	Kathmandu University, Nepal
Mr. Biraj Dhungana	Green Road Waste Management, Nepal
Dr. Krishna Prasad Shrestha	Kathmandu University, Nepal
Dr. Shailendra Kr. Jha	Kathmandu University, Nepal
Dr. Nawaraj Sanjel	EnergizeNepal, Nepal

Conference Publication Editors

Dr. Surendra Sujakhu	Editor-in-Chief	Kathmandu University, Nepal
Dr. Ranjit Shrestha	Editor	Kathmandu University, Nepal

IOP Conf. Series: Materials Science and Engineering **1314** (2024) 011001 doi:10.1088/1757-899X/1314/1/011001

Supporting Partners

Department of Mechanical Engineering, Kathmandu University, Nepal

University Grant Commission, Nepal

Institute of Engineering, Tribhuwan University, Nepal

Society of Mechanical Engineering Students

Automobile and Mechanical Engineering Students' Innovative Nexus

Society of Mechanical Engineers' Nepal

Society of Innovative Mechanical Engineering Students

Hosted by: Kathmandu University

Organized by: Association of Mechanical Engineering Students

IOP Conf. Series: Materials Science and Engineering **1314** (2024) 011002 doi:10.1088/1757-899X/1314/1/011002

Peer Review Statement

All papers published in this volume have been reviewed through processes administered by the Editors. Reviews were conducted by expert referees to the professional and scientific standards expected of a proceedings journal published by IOP Publishing.

- **Type of peer review:** Single Anonymous
- **Conference submission management system:** Morressier
- **Number of submissions received:** 36
- **Number of submissions sent for review:** 32
- **Number of submissions accepted:** 15
- **Acceptance Rate (Submissions Accepted / Submissions Received × 100):** 41.7
- **Average number of reviews per paper:** 1
- **Total number of reviewers involved:** 17
- **Contact person for queries:**
  Name: Dr. Surendra Sujakhu
  Email: surendra.sujakhu@ku.edu.np
  Affiliation: Department of Mechanical Engineering, Kathmandu University

IOP Conf. Series: Materials Science and Engineering **1314** (2024) 012001 doi:10.1088/1757-899X/1314/1/012001

Synthesis of dimethyl ether from biomass gasification: A simulation and techno-economic analysis

R Shrestha[1*], A Upreti[1] and N Solanki[1]

[1]Department of Chemical Science and Engineering, Kathmandu University, Dhulikhel, Nepal.

*Email: reshav.sth@gmail.com

Abstract. With ever-increasing concerns over environmental hazards, energy security, and, fleeting reserves of petroleum-based fuel, dimethyl ether (DME) has been under scrutiny as an alternative fuel, which is a high-efficiency compression ignition fuel with minimal NOx, SOx, and particulate matter emission. This paper explores the efficacy of biomass gasification as a method to synthesize dimethyl ether. The impact of steam and pure oxygen on syngas production has been studied, along with an analysis of how temperature affects the yield of DME. A process simulating software, ASPEN PlusV10.0, was used to simulate the synthesis process using the Langmuir-Hinshelwood-Hougen-Watson (LHHW) model. The direct (one-step) method has been studied where methanol synthesis and dehydration of methanol over a catalyst occur in a single plug flow reactor. Both steam and oxygen increased hydrogen production. Steam reduced CO and increased CO_2 production, whereas oxygen enhanced CO and decreased CO_2 production. From a biomass of 10,000 kg/hr, a yield of 4,583 kg/hr of dimethyl ether was achieved through the utilization of steam and oxygen as gasifying agents. A tentative economic analysis has been performed to understand the feasibility of the project. The CAPEX amounted to 64.7 million USD, while OPEX amounted to 28.9 million USD. It yielded an annual rate of return of 10.46%. The purity of DME was measured at 98.89%.

Keywords—dimethyl ether, gasification, methanol

1. Introduction

Fossil fuel has been one of the most important discoveries of mankind. It supplies 80 percent of the world's energy demand and has been powering homes for centuries. Although it has irrevocably shaped human civilization, its pivotal role in climate change, environmental pollution, and greenhouse emissions has made it subject to widespread controversy, and its consumption is still increasing steadily [1]. Biomass feedstocks, an organic derivative of residual plants and animals, play a pivotal role in developing alternatives to fossil fuels. Out of all renewable energy sources, only biomass provides the potential to generate liquid, carbon-neutral fuels [12].

Over the last twenty years, efforts have been made to investigate the potential of utilizing biomass as a promising resource for generating biofuels like dimethyl ether (DME)[2]. DME, a versatile compound, has gained attention due to its wide application, such as liquified petroleum gas (LPG), a second-generation fuel, due to its high cetane number, high-efficiency properties in compression ignition, and its minimal emissions of NOx, SOx, and particulate matter. DME can be produced from feedstocks like rice straw and sawdust [2,3,12]. Apart from biofuel, it can also be used as a refrigerant, Bukola et al. investigated the refrigeration effect of dimethyl ether, which performed better than existing refrigerant R134a, a global warming causing agent, and was considered an appropriate replacement for R134a refrigerators[13].

The conventional method involves dimethyl ether synthesis from syngas derived from natural gas. Syngas is a mixture of hydrogen and carbon monoxide at various ratios, which directly influence dimethyl ether yield. Biomass being carbon neutral and sustainable is a desirable source of producing syngas because its gasification produces clean synthetic fuel. One of the key factors in syngas composition is the gasifying agent. Air is frequently utilized as a gasification agent for biomass. However, the syngas produced from air gasification have a lower hydrogen content, which is not ideal for favorable dimethyl ether yield [14]. The hydrogen concentration and quality of syngas can be improved by using steam as a gasifying agent. The application of steam reduces residual tar formation, and the syngas produced in this manner have a higher heating value since this process creates more hydrogen than gasification with other agents [15]. Using air as the gasification agent is convenient and inexpensive. However, the high nitrogen content in the producer gas diminishes both its calorific value and the overall energy efficiency of the gasification process. Utilizing oxygen as a gasification agent can mitigate this issue; however, separation can incur high costs. Furthermore, it enhances oxidation reactions during the gasification process, resulting in a better yield of syngas[1,6,17]. Pengmei et al. [27] experiments showed biomass gasification using oxygen and steam significantly enhances hydrogen production compared to using air, nearly doubling the heating value of produced syngas in a downdraft gasifier. The gasification of biomass and reforming of producer gas are crucial phases in DME production, impacting both energy efficiency and costs. The gasification of biomass and reforming processes are key processes in DME production that have a great impact on energy efficiency and the cost of the entire process. Biomass gasification faces the issue of excess carbon in the form of CO_2 in the produced syngas. In the single-step synthesis reaction of DME from syngas, CO_2 is generated as a byproduct[18].

This study aims to demonstrate the feasibility of synthesizing DME directly from biomass gasification using both steam and pure oxygen as gasifying agents and aims to identify the optimal process conditions that maximize DME yield and purity. It aims to establish a relationship between the increasing amounts of gasifying agents and their effect on syngas production. The composition of syngas, which directly impacts the yield of dimethyl ether (DME) is examined. Additionally, it explores the effects of process parameters, such as temperature and reactor length on DME synthesis. An economic analysis is performed to assess the feasibility of the entire DME plant, providing insights into the evaluation of both the technical and economic aspects of the process.

2. Materials and Methods

2.1. Methodology

2.1.1 Feed Stock Selection

Douglas fir(pine) barks were selected as the biomass feedstock for the simulation based on ultimate and proximate analysis. The decision to opt for woody biomass was primarily based on its abundant availability, as well as its complete absence of sulfur and chlorine. If there had been notable levels of sulfur and chlorine, it would have required additional purification processes, like desulfurization and acid gas removal systems. This, in turn, would have raised the project's initial capital and ongoing operational costs.

Table 1 Basic properties of Douglas fir. [5]

Property	Value
Proximate analysis[wt% dry basis]	
Fixed Carbon	25.8
Volatile matter	73
Moisture	0
Ultimate analysis[wt% dry basis]	
Carbon	56.2
Hydrogen	5.9
Oxygen	36.7
Ash	1.2

Table 2 Input parameters for Aspen Simulation [5]

Property	Value
feedstock	Douglas Fir
stream class	MCINPSD[a]1
enthalpy	HCOALGEN[b]2
density	DCOALIGT[b]
Biomass flow rate	10000 kg h^{-1}
Steam flow rate	2250 kg h^{-1}
oxygen flow rate gasifier	1500 kg h^{-1}
temperature	1700 °C
inlet steam temperature	1200 °C
gasifier pressure	2MPa
[a]1 Standard Aspen Plus stream class	
[b]2 Standard Aspen Model	

2.1.2 DME Synthesis Method Selection

The traditional method for producing dimethyl ether, known as the indirect method, involves the dehydration of methanol and the synthesis of dimethyl ether in separate vessels. However, this indirect approach is more complicated as it requires high capital investment, and operating costs and includes multiple process stages. The primary drawback of this method is its limited conversion efficiency due to thermodynamic constraints in methanol synthesis[4]. Haldor Topsoe developed a technology that enables the concurrent execution of methanol synthesis and methanol dehydration within a single reactor from syngas, biomass, or natural gas. This approach, known as a one-step or direct synthesis, offers improved thermodynamic and economic advantages compared to the traditional two-step process[1,4,17]. Additionally, it requires a significantly lower initial capital investment as it has fewer stages in comparison to the two-step method. Teerana et al.[14] experiments on air-steam gasification with CO_2 utilization conclude that DME yield in the direct process was significantly higher than in the indirect method. Furthermore, the indirect approach for biomass gasification and DME synthesis consumes significantly more energy and destroys exergy than the direct system.

In this study, we employed the direct synthesis method for the production of dimethyl ether from syngas. This approach integrates methanol synthesis and methanol dehydration into a single reactor, utilizing a dual-function catalyst system. The direct synthesis process offers thermodynamic and economic advantages over the traditional two-step process by reducing the number of process stages and enhancing overall conversion efficiency.

2.1.3 Biomass Gasification Unit

Biomass gasification is a thermochemical process that converts carbon-rich biomass into gaseous products. Unlike traditional coal gasification, which occurs at higher temperatures, biomass gasification operates at around 900°C due to the nature of the biomass. The high volatile content and the presence of intrinsic catalytic metals such as potassium and calcium in biomass enhance its reactivity. Moreover, biomass is environmentally friendly, contributing no net greenhouse gas emissions and having low sulfur and nitrogen content, making it a cleaner alternative to fossil fuels. The gasification process converts the elements such as C, H, 0_2, Cl, S, and N_2 present in biomass into syngas, a mixture of H_2, CO, CO_2, and CH_4 at various ratios, ash, and char by the combination of complex homogeneous and heterogeneous reaction[19]. Syngas is then used as feed to the DME reactor.

For this study, we used an entrained flow gasifier due to the absence of tar formation, and it generally uses oxygen as feed. In an entrained flow gasifier, the residence duration is usually a few seconds or tens of seconds with high carbon conversion efficiencies (98-99.5%) and an extremely turbulent flow. These gasifiers have a short residence period; therefore, to achieve high carbon conversion, gasifiers run at temperatures over feedstock's slagging point and employ oxygen rather than air, which allows for rapid feed conversion and high throughput. The entrained-flow gasifier can process a variety of feedstock, including biomass and coal, and generate clean, tax-free syngas [20].

2.1.4 DME Synthesis Unit: After syngas purification, the syngas were conveyed to the DME synthesis reactor. In the direct DME synthesis method, a bifunctional catalyst is used where methanol synthesis and methanol dehydration take place at the same time. The reactor is modeled as a plug flow reactor due to high conversion efficiency, uniform residence time, and enhanced mass transfer.

2.1.5. DME Purification Unit: Subsequently, DME was isolated from the gases generated by the DME synthesis reactor and was further purified to attain the desired level of DME purity by a series of distillations.

2.1.6 Process Description

Before entering the gasification chamber, biomass undergoes pretreatment involving pulverization to a size range of 200-300 μm through a primary and secondary system, followed by drying to minimize moisture content. The introduction of biomass into the high-temperature (1700°C) and high-pressure (20 bar) gasification chamber is facilitated by a screw conveyor. Steam and oxygen serve as the gasifying agents in this process. The utilization of steam aids in the generation of H_2 by promoting the water gas shift reaction, while oxygen ensures the effective combustion of char by providing the necessary heat for gasification. The gasifier's output is directed to a cyclone, where solids are separated from the gas streams. Given the absence of sulfur and chlorine in the initial feed, no supplementary purification is required. The high-temperature, purified syngas are subsequently cooled using a heat exchanger, with water at 15°C employed as a cooling agent.

The process requires a pressure of 40 bar, leading the cooled syngas to be sent to a compressor. Subsequently, the cooled syngas, now at an elevated pressure, are directed into the DME Synthesis reactor, where methanol formation and methanol dehydration take place. These reactions are facilitated by a bifunctional catalyst composed of $CuO/ZnO/Al_2O_3$ and γ-Al_2O_3. Given the highly exothermic and reversible nature of these reactions, an isothermal reactor is favored over an adiabatic one.

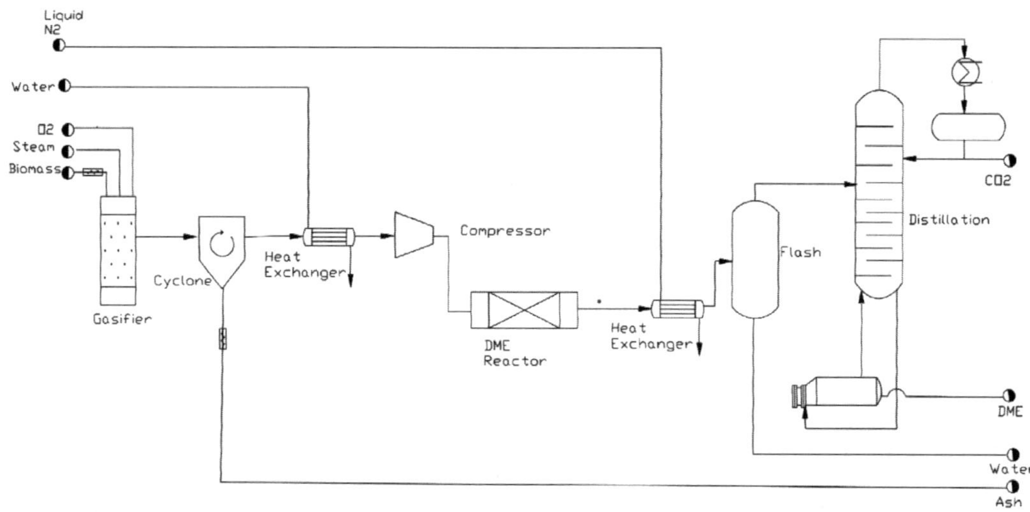

Figure 1. Process flow diagram of DME plant

This process turns out to be beneficial as it yields superior results by avoiding thermodynamic constraints on conversion. The DME Synthesis reactor operates at approximately 250°C and 40 bar pressure. Considering the boiling points, CO_2 has a boiling point of -78°C, while DME boils at -24°C at 1 bar. DME, conveniently, can be stored in liquid form at standard temperature and elevated pressure (5 bar). The product from the DME reactor undergoes cooling in a heat exchanger to approximately -40°C, achieved by using liquid nitrogen as a coolant. During this cooling process, liquid phase separation occurs in a flash unit, leading to a pressure drop of 25 bar as water and a small amount of methanol are separated. The vapor from the flash unit is then directed into a cryogenic distillation column operating at 5 bar, featuring 18 trays and a reflux ratio of 3.6. This setup is designed to attain a high DME purity of up to 98.89%. The resulting liquid DME is stored in a storage tank at a pressure of 5 bar.

The top product from the distillation column consists of uncondensed CO_2 at around -60°C. This CO_2 is recycled as a cooling agent to lower the temperature of water from 25°C to 15°C, which is then used to further cool the hot syngas from the cyclone. The remaining unreacted hot gas is released into the atmosphere through a chimney.

3. Aspen Plus Simulation

The simulation was conducted using Aspen Plus, a software capable of accurately modeling steady-state fluid flow processes. In this modeling, BIOMASS and ASH were categorized as nonconventional compounds, while carbon was designated as a solid component. To determine the enthalpy and density of non-conventional compounds like biomass and ash, the HCOALGEN and DCOALIGT models were used. The gasifier was simulated using two different reactors (RYield and RPlug). Firstly, the BIOMASS stream was introduced into the DECOMP (RYield) block. Within this stage, the mass flow rates of individual components were set by establishing a yield distribution, which was determined from the ultimate analysis data of Douglas fir. It's important to note that this isn't a separate reactor but an essential component integrated into the gasification reactor system. The purpose of the RYield reactor was to transform nonconventional solids into gaseous components and ash before they were directed into the Plug Flow reactor. The Peng-Robinson thermodynamic package was used for gasification as it consists of a temperature-dependent variable alpha that can calculate vapor pressure and other thermodynamic properties at higher temperatures more accurately. The gasifier(RPlug) converts the decomposed biomass into syngas. The syngas is sent into a solid gas separator, cyclone(SSplit). Now the syngas(gas only) is directed into a heat exchanger (HeatX) and it cools down the gas to a moderate 250°C. The cooled syngas is sent to a plug flow reactor(RPlug). The Soave-Redlich-Kwong(SRK) property method was used for the DME synthesis reactor. The gas combination exiting the DME synthesis reactor underwent cooling via the HX2 (HeatX) block, resulting in the removal of non-condensable gases through a FLASH (Flash) separator. Ultimately, the DME product was acquired as the lower output, while liquid CO_2 was collected as the upper yield from a distillation column (Distl). The top product from the distillation tower is recycled back as coolant to HX3. The optimum values of different equipment parameters were obtained from Aspen plus by hit and trial method and different sensitivity analysis. The desired H_2/CO molar ratio in the gasifier outlet was 2.0.

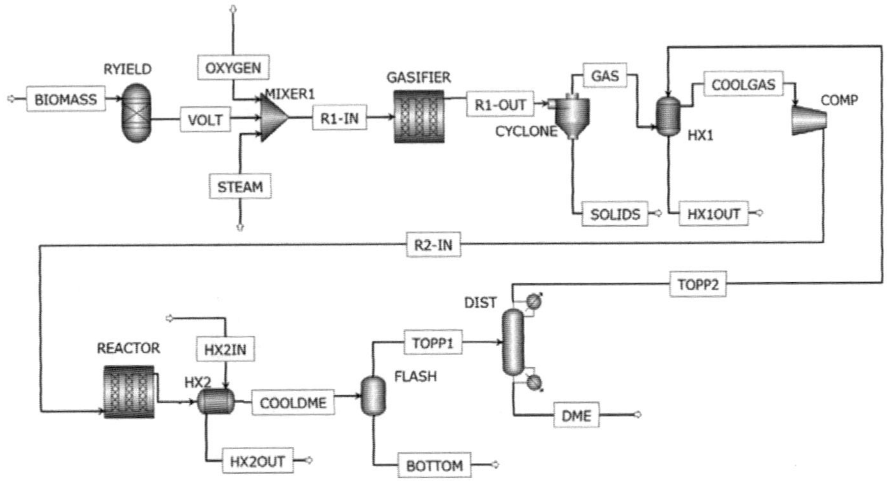

Figure 2. Aspen flowsheet of single-step DME production through steam/oxygen biomass gasification.

4. Reaction Involved

4.1Gasification

Since gasification is a complex process, there are many reactions occurring in series. The kinetics model used for gasification is driven by a simple power law. There are a series of heterogeneous and homogeneous reactions involved in the gasification process. Some of the major combustion and gasification reactions considered in this process were:

4.1.1. Heterogeneous reactions [6]

The char produced in the gasification is consumed by heterogeneous reactions.

$$C + H_2O \leftrightarrow CO + H_2 \qquad (R1)$$
$$C + CO_2 \leftrightarrow 2CO \qquad (R2)$$
$$C + 2H_2 \leftrightarrow CH_4 \qquad (R3)$$
$$C + 0.5O_2 \rightarrow CO \qquad (R4)$$

The initial three equations are formulated as equilibrium reactions that consider the influence of reverse reactions. The rate expressions for (R-1), (R2), (R-3), and (R-4) are summarized in Table 3.

4.1.2. Homogeneous reactions[6]

In this current study, the homogeneous gas-phase reactions include combustion reactions of CO, H_2, CH, and C_2 H_4, the water-gas shift reaction, and the methane steam reforming reaction.

$$H_2 + 0.5O_2 \rightarrow H_2O \qquad (R5)$$
$$CO + H_2O \leftrightarrow CO_2 + H_2 \qquad (R6)$$
$$CH_4 + 2O_2 \rightarrow CO_2 + 2H_2O \qquad (R7)$$
$$CH_4 + H2O \rightarrow CO + 3H_2 \qquad (R8)$$
$$CO + 0.5O_2 \rightarrow CO_2 \qquad (R9)$$
$$CH_4 + H_2O \rightarrow CO + 3H_2 \qquad (R10)$$

Table 3. Rate expression for reactions [6]

S.N.	Reaction	Rate Law
R1	$C + H_2O = CO + H_2$	$5.714*T*exp(-1.29706 \times 10^5/R.T).[H_2O]$
R2	$C + CO_2 = 2C0$	$589*T*exp(-2.22829 \times 10^5 /R.T).[CO_2]$
R3	$C + 2H_2 = CH_4$	$0.00342*T*exp(-13 \times 10^3/R.T).[H_2]$
R4	$C + 0.5O_2 = C0$	$3.7*10^{10}*T*exp(-1.5*10^5/R.T).[O_2]$
R5	$CO + H_2O = CO_2 + H_2$	$7.68 *10^{10}*exp(-3.046 \times 10^5 /R.T).[CO]^{0.5} .[H_2O]$
R6	$CO + 0.5O_2 = CO_2$	$5.62*10^{12}*exp(-1.33 \times 10^5/R.T).[O2]^{0.5}.[CO]$
R7	$CH4 + 2O_2 = CO_2 + 2H_2O$	$3.552*10^{11}*exp(-1.30529 \times 10^5/R.T).[O_2].[CH_4]$
R8	$CH_4 + H_2O = CO + 3H_2$	$3 *10^5* exp(-1.25 \times 10^5/R.T) [CH_4].[H_2O]$
R9	$CO + 0.5O_2 = CO_2$	$5.62*10^{12}*exp(-1.33 \times 10^5/R.T).[O2]^{0.5}.[CO]$
R10	$CH_4 + H_2O = CO + 3H_2$	$3 *10^5* exp(-1.25 \times 10^5/R.T) [CH_4].[H_2O]$

4.2 Dimethyl Ether[7]

Similarly, the DME synthesis was carried out in the presence of a bifunctional $CuO/ZnO/Al_2O_3$ γ-Al_2O_3 as a catalyst and the Langmuir-Hinshelwood-Hougen-Watson (LHHW) model was used as a kinetic model since it involves a heterogeneous reaction between solid and gas phase. The reactions involved in DME synthesis are:

4.2.1 Methanol Synthesis Reaction

$CO+2H_2 \rightarrow CH_3OH$ (R11)................ ΔH=-128.2 kJ/mol
$CO_2+3H_2 \rightarrow CH_3OH+H_2O$ (R12)............. ...ΔH=-131.0 kJ/mol

4.2.2. Methanol Dehydration

$2CH_3OH \rightarrow CH_3OCH_3$ (*R*13)................ ΔH= -238.7 kJ/mol.

4.2.3. Water Shift Gas Reaction

$CO + H_2O \rightarrow CO_2 + H_2$ (R14).............. ΔH= -41.0 kJ/mol.

The rate equation is given as:[7,8]

$$r_{CO} = \frac{k_1 f_{CO} f_{H_2}^2 - \frac{k_1 f_{CH_3OH}}{K_{f_1}}}{\left(1 + K_{CO} f_{CO} + K_{CO_2} f_{CO_2} + K_{H_2} f_{H_2}\right)^3}$$

$$r_{CO_2} = \frac{k_2 f_{CO_2} f_{H_2}^3 - \frac{k_2 f_{CH_3OH_{H_2O}}}{K_{f_2}}}{\left(1 + K_{CO} f_{CO} + K_{CO_2} f_{CO_2} + K_{H_2} f_{H_2}\right)^4}$$

$$r_{DME} = \frac{k_3 f_{CH_3OH} - \frac{k_3 f_{DME} f_{H_2O}}{K_{f_3}^2 f_{CH_3OH}}}{\left(1 + \sqrt{K_{CH_3OH} f_{CH_3OH}}\right)^2}$$

$$r_{WGS} = \frac{k_4 f_{H_2O} - \frac{k_4 f_{CO_2} f_{H_2}}{K_{f_4} f_{CO}}}{1 + K_{CO} f_{CO} + K_{CO_2} f_{CO_2} + \sqrt{K_{H_2} f_{H_2}}}$$

Table 4 Kinetic Parameters[7,8]

Parameter = A(i)exp (B(i)/ RT)	**A(i)**	**B(ii)**
k_1	7380	−54,307
k_2	5059	−57,515
k_3	1062	−43,473
k_4	7.3976	−20,436
K_{CO}	3.934×10^{-6}	37,373
K_{CO2}	1.858×10^{-6}	53,795
K_{H2}	0.6716	-6476
K_{CH3OH}	3.480×10^{-6}	54,689

5. Results and discussion

5.1 Effect of steam as a gasifying agent in Gasifier

Steam affects the gasification process by influencing hydrogen production, gas composition, and gas yield[9]. Figure 3 shows the mass flow rate of syngas yields versus steam. When the steam flow rate was increased by 2 folds from 1000 kg/hr, the H_2 and CO_2 production increased by 178.8% and 140% respectively while at the same time CO production decreased by 23.8%. In their research on biomass gasification using air and steam mixtures, Campoy et al.[24] and Tian et al.[25] also noted an increase in H_2 and CO_2 concentrations, accompanied by a decrease in CO concentrations under similar conditions. The higher steam flow rate tends to facilitate more hydrogen production. It is assumed that steam promotes the water shift gas reaction (R5), which converts CO and steam into H_2 and CO_2[5]. This reaction is endothermic, and its equilibrium shifts towards the production of hydrogen and carbon dioxide with increasing steam and temperature. Endothermic reactions (R6), (R7), and (R10) contributed to the augmentation of H_2, which is a direct outcome of increased temperature caused by steam flow rate.

The initial incomplete combustion of feedstock is assumed to be the reason for significant CO formation. However, when the steam flow rate increases, the water-gas shift reaction is more prevalent, converting CO into CO_2 and H_2. The increase in CO_2 can also be explained by the same reaction which is favored by a higher flow rate. Another possible explanation might be an intensification of the reaction(R9) which is a result of an increment in temperature by steam.

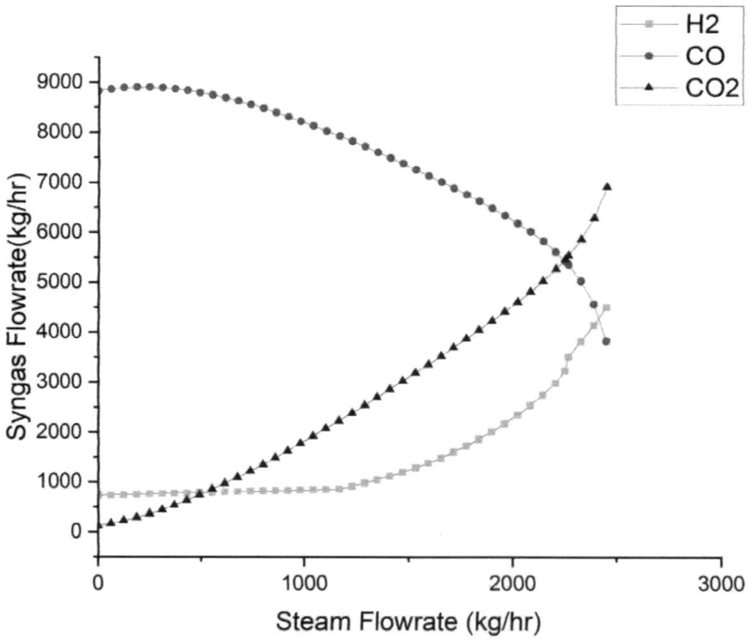

Figure 3. Syngas Composition Flowrate vs. Steam Flowrate in Biomass Gasifier

5.2. Effect of oxygen as a gasifying agent

Oxygen plays a crucial role in enhancing gasification efficiency and helps in the combustion of feedstock that helps in generating heat within the gasifier necessary for other gasification reactions to occur. Figure 4 shows an increase in the production of H_2 reaching a maximum of 5400 kg/hr when the oxygen flow rate is increased. A similar trend is observed in CO while CO_2 production declined. These results are in agreement with the results presented by Wang et al.[23] where H_2 content in syngas obtained from pine sawdust increases from 16% to 36.45% when oxygen concentration is increased from 21 to 99 v%. This increase can be anticipated because the local temperature at the surface of the reacting particles and their boundary layer is likely to rise with higher O_2 concentration in the gasification agent, due to the heat released from intensified exothermic reactions. As a result, the higher temperature would favor endothermic reactions, such as water-gas (R2) and Boudouard (R2), reforming, and tar cracking reactions. This would promote tar destruction and char conversion, thereby increasing H_2[21].

Boudouard reaction(R2) may help explain the increase in CO and decrease in CO_2. This reaction consumes CO_2 and produces CO. As the oxygen mass flow rate increases, more CO_2 is produced from the combustion of carbon. However, the increased temperature also favors the Boudouard reaction to the right, driving the conversion of CO_2 to CO. This explains why the CO mass flow rate increases with increasing oxygen flow rate. The mass flow rate of CO_2 declines after reaching a point. Water-gas shift reaction(R5) consumes CO and produces CO_2 and H_2. With increasing oxygen flow rate, more H_2O is also produced from the combustion of hydrogen in the feedstock. This leads to an increase in the water-gas shift reaction, consuming CO and producing CO_2 and H_2. However, the Boudouard reaction counteracts this effect to by consuming CO_2 and producing CO. The net effect of these two competing reactions is that the CO mass flow rate increases initially with increasing oxygen flow rate due to the dominance of the Boudouard reaction.

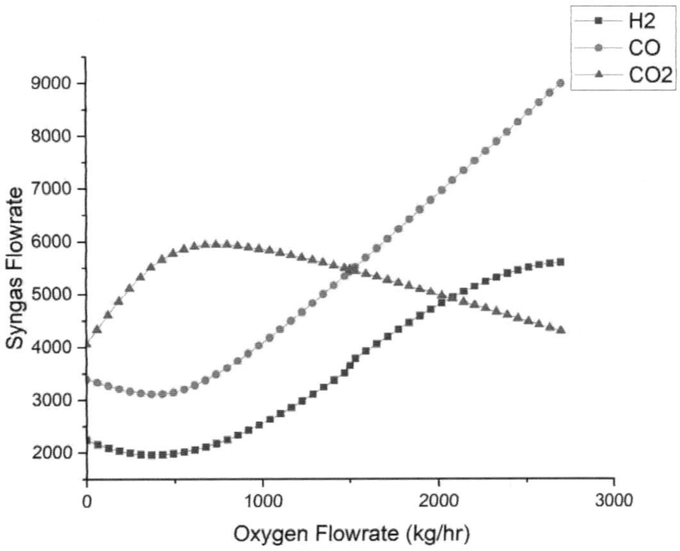

Figure 4. Syngas Composition Flowrate vs. Oxygen Flowrate in Biomass Gasifier

5.3. Effect of temperature on DME synthesis reactor

Temperature has a significant effect on DME production. The results of the simulation study for the effect of temperature are presented in Figure 5. A steady increase can be observed between DME flow rate and temperature, as can be observed in Figure 5a. The DME yield increases steadily up to 4500 kg/hr from 500°C to 540°C until it plateaus due to thermal equilibrium constraints. It reaches maximum conversion at 545°C. Nirvana et al.[22] experiments revealed higher temperatures result in increased DME yield and selectivity, irrespective of the feed composition. This increase can be explained by the temperature dependence of the rate constant. The rate of the methanol dehydration reaction, which is an equilibrium reaction, increases with an increase in temperature. Higher temperatures provide more thermal energy to reactant molecules, allowing them to overcome the activation energy barrier more easily. Higher temperatures influence the interactions among methanol molecules, raising the chances of successful reactions and the production of dimethyl ether and water. It can also be explained by Le Chatelier's principle, that increasing the temperature favors the endothermic direction of the reaction, which is the formation of DME and water. Therefore, raising the temperature tends to shift the equilibrium towards higher DME yields.

In Figure 5b, CO_2 production is also increasing at elevated temperatures, reaching 5100 kg/hr at 540°C. The increased yield of CO_2 can be explained by the water shift gas reaction, where low temperature favors the product side. Higher temperatures generally increase the rates of all reactions, but the water-gas shift reaction may have a lower activation energy compared to the methanol synthesis and dehydration reactions. This implies that at elevated temperatures, the rate at which CO_2 forms can accelerate more quickly than the rate at which DME forms. In Figure 5c, CH_3OH is being effectively consumed at methanol dehydration and other side reactions to yield dimethyl ether.

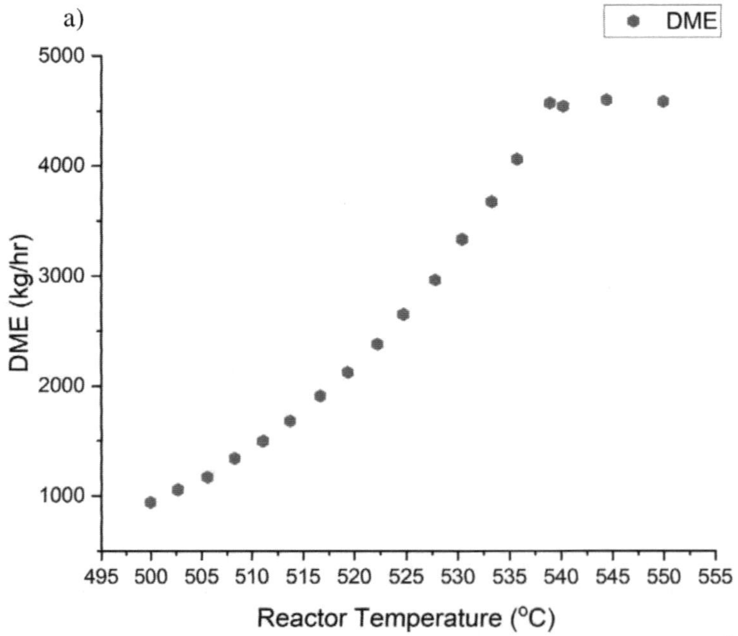

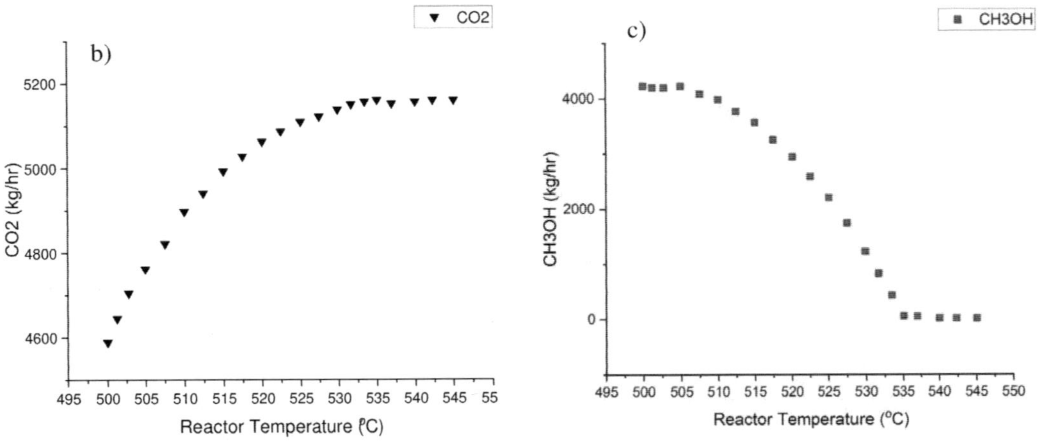

Figure 5. Effect of increasing temperature on yield of a)Dimethyl Ether. b)Carbon Dioxide c)Methanol

5.4. Temperature profile along the length of the gasifier

Temperature plays a crucial role in gasifiers by strongly influencing the speed of gasification reactions. From Figure 6, it can be seen that there is a step increase in temperature at the beginning of the reactor length. It can be assumed that this section of the reactor is where most of the exothermic reaction occurs, releasing significant heat. This is where maximum combustion of biomass with oxygen is likely to happen. The exothermic reactions reach their thermodynamic equilibrium at around 1m attaining a peak temperature of 1800K. Beyond the peak, the temperature falls gradually along the length of the reactor. This might be due to heat consumed by endothermic reactions water-shift gas reactions (R5), and Boudouard reaction(R2). Towards the end of the reactor length, the temperature seems to stabilize. This suggests that the reactions have reached a steady state. The heat generation and consumption are balanced and it attains a constant temperature. For a plug flow reactor, increasing the temperature can lead to a shorter required reactor length. This occurs because elevated temperatures frequently lead to increased reaction rates, enabling equivalent conversion levels to be reached over a shorter distance.

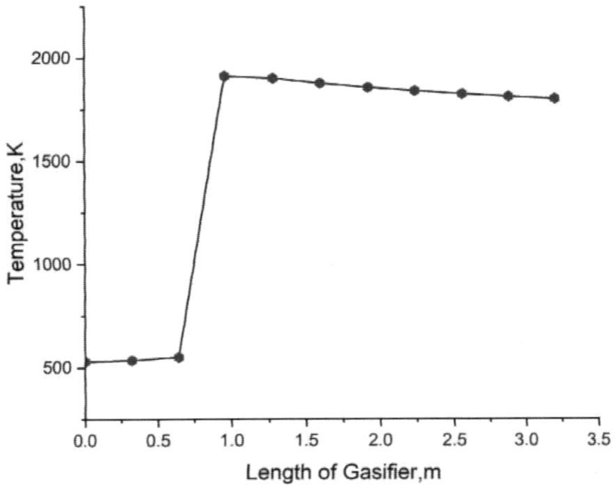

Figure 6. Temperature profile along the length of the gasifier

5.5. Economic Analysis

The economic analysis is an important factor in assessing the feasibility of a project. The profitability of a company depends on the economics of its operations. Likewise, for a project to be considered successful, it should reach its breakeven point within an appropriate timeframe. The detailed calculations are presented in our thesis work. The DME production capacity of the plant is calculated to be 4583 kg/hr for 300 days. The steam flow rate was 2250 kg/hr, and the oxygen flow rate was 1500 kg/hr with a purity of 97%. This process is extremely energy-intensive, with temperatures ranging from approximately 1700°C in the gasifier to -60°C in the condenser. The CAPEX, total capital expense of all machinery, land, buildings, etc, amounted to approximately 64.7 million USD, and the OPEX, annual cost of operating the plant was around 28.9 million USD. The cost of utilities such as steam, oxygen, and nitrogen were 0.03, 0.05, and 0.0074 USD/kg respectively. The plant yielded an annual rate of return of 10.46%. The utilization of liquid nitrogen as a coolant significantly contributed to the increase in operating costs. The purity of the obtained dimethyl ether (DME) was measured at 98.89%. Purity was measured based on the final stream composition in Aspen PlusV10.0[28]. Based on current market prices, high-purity DME of this kind is valued at about 1.15 USD per kg as of 2023[11]. The cost of producing DME is calculated at 0.87 $/kg, which equates to 0.76 $/l, considering a density of 0.873 kg/l. After deducting a 25% tax, the net profit obtained from this process amounted to 6.7 million USD. The project has a breakeven period of 8 years.[28]

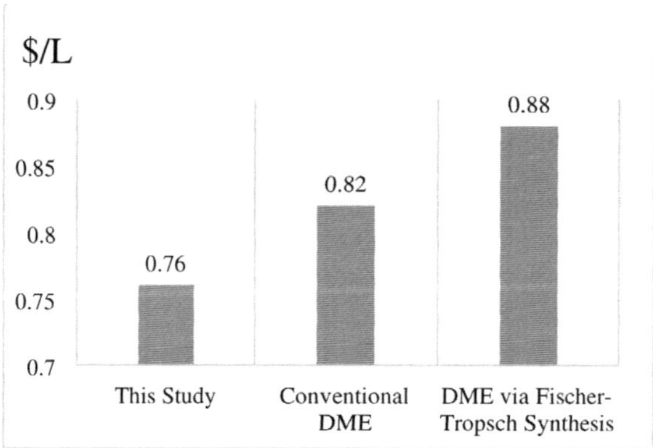

Figure 7. Comparison of DME from different sources, Conventional DME[10], DME via Fischer Tropsch Synthesis[10]

6. Conclusions

In conclusion, the production of dimethyl ether from Douglas fir biomass gasification presents a promising avenue for the sustainable production of a clean-burning fuel. Through this study, we simulated and concluded that biomass gasification can effectively convert woody biomass using steam/pure oxygen as a gasifying agent into syngas containing high levels of hydrogen and carbon monoxide, which can then be synthesized into DME via a catalytic one-step process from existing technologies. This renewable fuel has the potential to replace conventional fossil fuels in various applications, including transportation and heating, thereby reducing greenhouse gas emissions and promoting energy security. The study highlights the influence of gasifying agents and temperature on the efficiency and product composition in biomass gasification and DME synthesis. The process was modeled on simulating software, Aspen PlusV10.0, and results reveal, that increasing steam flow rate enhanced hydrogen and carbon dioxide yield while reducing CO yield asserting previous findings. Similarly, the introduction of pure oxygen also improved the quality of syngas. Higher conversion of DME was obtained which may be attributed to significant CO production when O_2 was introduced. This is because the higher availability of CO facilitates methanol synthesis reaction(R11). Temperature seems to be a contributing factor to the high yield of DME up to a certain point. The highest conversion of DME was achieved at 545°C.

This approach was also feasible from an economic standpoint even using utilities like pure oxygen and liquid nitrogen as coolant. Through 10000kg/hr of pulverized and dried woody biomass, we have produced around 4583 kg/hr of DME with purity of 98.89%. The CAPEX was calculated at 64.7M USD and OPEX was 28.9M USD per year. The production rate was 0.87$/kg DME and against a market price of 1.13$/kg[11] it was able to generate a total revenue of 37.9M USD with the rate of return of 10.46%. The payback period was 8 years and hence through different economic parameters, it can be concluded that the project is investable and profitable. However, the production of DME from woody biomass gasification still faces several challenges, including issues related to catalyst stability, scalability, high energy requirement, and economic viability.

Overall, this study contributes to the knowledge of the production of renewable fuels from biomass and provides valuable insights into the potential of DME as a sustainable energy carrier. The use of steam and oxygen contributed to high conversion and highly pure DME. Further research can be done to explore the use of different catalysts and reaction conditions for DME synthesis to achieve higher yields and selectivity. More studies can be carried out to investigate the possibility of using different feedstocks for gasification to produce syngas for DME synthesis.

7. References

[1] Arcoumanis, C., Bae, C., Crookes, R. and Kinoshita, E., 2008. The potential of di-methyl ether (DME) as an alternative fuel for compression-ignition engines: A review. *Fuel*, *87*(7), pp.1014-1030.

[2] Silalertruksa, T., Gheewala, S.H., Sagisaka, M. and Yamaguchi, K., 2013. Life cycle GHG analysis of rice straw bio-DME production and application in Thailand. *Applied Energy*, *112*, pp.560-567.

[3] Azizi, Z., Rezaeimanesh, M., Tohidian, T. and Rahimpour, M.R., 2014. Dimethyl ether: A review of technologies and production challenges. *Chemical Engineering and Processing: Process Intensification*, *82*, pp.150-172.

[4] Ju, F., Chen, H., Ding, X., Yang, H., Wang, X., Zhang, S. and Dai, Z., 2009. Process simulation of single-step dimethyl ether production via biomass gasification. *Biotechnology advances*, *27*(5), pp.599-605.

[5] Parvez, A.M., Mujtaba, I.M., Hall, P., Lester, E.H. and Wu, T., 2016. Synthesis of Bio-Dimethyl Ether Based on Carbon Dioxide-Enhanced Gasification of Biomass: Process Simulation Using Aspen Plus. *Energy Technology*, *4*(4), pp.526-535.

[6] Xie, J., Zhong, W., Jin, B., Shao, Y. and Liu, H., 2012. Simulation on gasification of forestry residues in fluidized beds by Eulerian–Lagrangian approach. *Bioresource technology*, *121*, pp.36-46.

[7] Polsen, C., Narataruksa, P., Hunpinyo, P. and Prapainainar, C., 2020. Simulation of single-step dimethyl ether synthesis from syngas. *Energy reports*, *6*, pp.516-520.

[8] Nie, Z., Liu, H., Liu, D., Ying, W. and Fang, D., 2005. Intrinsic kinetics of dimethyl ether synthesis from syngas. J. Nat. Gas Chem, 14(1), pp.22-2

[9] Baláš, M., Lisý, M., Pospíšil, J., 2016. Steam Biomass Gasification - Effect of Temperature. AMM.

[10]Trippe, F., Fröhling, M., Schultmann, F., Stahl, R., Henrich, E. and Dalai, A., 2013. Comprehensive techno-economic assessment of dimethyl ether (DME) synthesis and Fischer–Tropsch synthesis as alternative process steps within biomass-to-liquid production. Fuel processing technology, 106, pp.577-586.

[11]Businessanalytiq, "Dimethyl Ether price index,". [Online]. Available: Dimethyl Ether price index - businessanalytiq.

[12]Chang, J., Fu, Y. and Luo, Z., 2012. Experimental study for dimethyl ether production from biomass gasification and simulation on dimethyl ether production. Biomass and bioenergy, 39, pp.67-72.

[13]Bolaji, B.O., Oyelaran, O.A., Abiala, I.O., Ogundana, T.O. and Amosun, S.T., 2021. Energy and thermal conductivity assessment of dimethyl-ether and its azeotropic mixtures as alternative low global warming potential refrigerants in a refrigeration system. Environmental and Climate Technologies, 25(1), pp.12-28.

[14]Nakyai, T., Patcharavorachot, Y., Arpornwichanop, A. and Saebea, D., 2020. Comparative exergoeconomic analysis of indirect and direct bio-dimethyl ether syntheses based on airsteam biomass gasification with CO2 utilization. *Energy*, *209*, p.118332.

[15]Alaedini, A.H., Tourani, H.K. and Saidi, M., 2023. A review of waste-to-hydrogen conversion technologies for solid oxide fuel cell (SOFC) applications: Aspect of gasification process and catalyst development. *Journal of Environmental Management*, *329*, p.117077.

[16]Xu, Chunbao & Liao, Baoqiang & Pang, Shusheng & Nazari, Laleh & Mahmood, Nubla & Tushar, Mohammad & Dutta, Animesh & Ray, Madhumita. (2018). 1.19 Biomass Energy. 10.1016/B978-0-12-809597-3.00121-8.

[17]Fleisch, T.H., Basu, A., Gradassi, M.J. and Masin, J.G., 1997. Dimethyl ether: A fuel for the 21st century. In *studies in surface science and catalysis* (Vol. 107, pp. 117-125). Elsevier.

[18]Chang, J., Fu, Y. and Luo, Z., 2012. Experimental study for dimethyl ether production from biomass gasification and simulation on dimethyl ether production. *Biomass and bioenergy*, *39*, pp.67-72.

[19]Yang, H. and Chen, H., 2015. Biomass gasification for synthetic liquid fuel production. In *Gasification for synthetic fuel production* (pp. 241-275). Woodhead Publishing.

[20]J. Phillips, "Different Types of Gasifi ers and Their Integration with Gas Turbines," National Energy Technology Laboratory. [Online]. Available: GTHandbook-final-edits-new.indd (doe.gov) . [Accessed:Jul. 8,2024]

[21]Gomes, H.G., Matos, M.A. and Tarelho, L.A., 2023. Influence of Oxygen/Steam Addition on the Quality of Producer Gas during Direct (Air) Gasification of Residual Forest Biomass. *Energies*, *16*(5), p.2427.

[22]Otalvaro, N.D., Sogne, G., Delgado, K.H., Wild, S., Pitter, S. and Sauer, J., 2021. Kinetics of the direct DME synthesis from CO 2 rich syngas under variation of the CZA-to-γ-Al 2 O 3 ratio of a mixed catalyst bed. *Rsc Advances*, *11*(40), pp.24556-24569.

[23]Wang, Z., He, T., Qin, J., Wu, J., Li, J., Zi, Z., Liu, G., Wu, J. and Sun, L., 2015. Gasification of biomass with oxygen-enriched air in a pilot scale two-stage gasifier. *Fuel*, *150*, pp.386-393.

[24]Campoy, M., Gomez-Barea, A., Vidal, F.B. and Ollero, P., 2009. Air–steam gasification of biomass in a fluidised bed: Process optimisation by enriched air. *Fuel Processing Technology*, *90*(5), pp.677-685.

[25]Tian, Y., Zhou, X., Lin, S., Ji, X., Bai, J. and Xu, M., 2018. Syngas production from air-steam gasification of biomass with natural catalysts. *Science of the Total Environment*, *645*, pp.518523.

[26]Nakyai, T., Patcharavorachot, Y., Arpornwichanop, A. and Saebea, D., 2020. Comparative exergoeconomic analysis of indirect and direct bio-dimethyl ether syntheses based on airsteam biomass gasification with CO2 utilization. *Energy*, *209*, p.118332.

[27]Lv, P., Yuan, Z., Ma, L., Wu, C., Chen, Y. and Zhu, J., 2007. Hydrogen-rich gas production from biomass air and oxygen/steam gasification in a downdraft gasifier. *Renewable energy*, *32*(13), pp.2173-2185.

[28]R Shrestha, A Upreti and N Solanki 2023 Synthesis Of Dimethyl Ether From Biomass Gasification *Bachelor* Thesis Kathmandu University, Dhulikhel.

IOP Conf. Series: Materials Science and Engineering **1314** (2024) 012002 doi:10.1088/1757-899X/1314/1/012002

An Experimental and Numerical Investigation on Fatigue Life Cycle of Leaf Spring

Narayan Guragain[1], Rodan Bista[1], Piyush Nepal[1], Birat Timilsina[1], Surendra Sujakhu*[1]

[1]Department of Mechanical Engineering, Kathmandu University, Nepal

E-mail: surendra.sujakhu@ku.edu.np

Abstract. Leaf spring is the key component of the vehicle suspension system that affects the mechanical behavior of vehicle vibration, stability, damping, and other aspects like comfort. However, depending upon the geography and road conditions, its fatigue life may vary. In Nepal, public vehicles operate mainly on paved roads and gravel roads under variable loading conditions. The road tested in this paper includes an urban road ranging 45km from Ratnapark to Sundarijal. The road consists of paved surfaces and gravel roads with varying degrees of slopes, potholes, and cracks. This paper discusses the fatigue life of a leaf spring in a public bus (TATA-709) operating on an urban road. Miner's rule is used for the experimental analysis. Miner's rule requires real-life data for the number of stress cycles. The data is obtained from the public bus using the data collection module. The number of repeated stress cycles was computed using the Rainflow cycle counting method and, the fatigue life is calculated implementing Miner's rule. To validate the experimental results, fatigue simulation is carried out in ANSYS. The stress history data obtained from the experiment is fed into the fatigue model in ANSYS. The result shows that the number of cycles to failure obtained from ANSYS exceeds the number of cycles of failure obtained from Miner's rule by 5%. This discrepancy indicates the need to consider further factors beyond this study, such as rusting, creep, and pre-stress, which considerably change the material properties and fatigue. This study sets a method to determine the fatigue life cycle of a leaf spring in various road and loading conditions both experimentally and numerically, to avoid potential damage due to failure of the suspension system in a public vehicle.

1. Introduction

The leaf spring affects the mechanical behaviors of a vehicle. It affects the vibration, stability, damping, and other aspects such as comfort. It is used to absorb shock and vibrations in heavy commercial vehicles. Thus, leaf spring experiences a large number of fatigue cycles reducing its life span. The number of jerks and fatigue experienced depends upon the road conditions, geographical terrain, climate, and temperature [1]. In rail and suspension systems for light trucks, heavy-duty vehicles, and other autos, leaf springs are primarily utilized to absorb shock loads. It handles driving torque, brake torque, side loads, lateral loads, and longitudinal loads [2]. Particularly in materials subjected to cyclic stresses, fatigue damage develops and eventually results in rupture [5]. Therefore, it is a crucial factor to consider when designing automobile structures, and aircraft structures. High-cycle fatigue, low-cycle fatigue, and studies of fatigue fracture formation are just a few examples of the many varied challenges and methodologies that are part of the large scientific and technology field known as fatigue design [6].

IOP Conf. Series: Materials Science and Engineering **1314** (2024) 012002 doi:10.1088/1757-899X/1314/1/012002

Therefore, lifetime calculations utilizing a cumulative damage law are often the basis for the fatigue design of engineered machine components and structures. The present case studies the fatigue life of leaf springs in the context of Nepal.

Various studies are carried out to estimate the life cycle of leaf springs. Most of the studies involved the initial calculation of fatigue life to determine the life cycle. These are mostly carried out by FEM (Finite element method) analysis and other methods. Kumar et al. [2] explain the static and fatigue study of steel leaf springs and composite multi-leaf springs composed of glass fiber-reinforced polymer utilizing life data analysis. In this research work, static analysis for various leaf springs is carried out using ANSYS, and the fatigue life of various composite leaf springs is calculated. Obtained deflection and von Mises stress are compared with theoretical results. The results are in the error proximity of 8-9%. Bhushan, et al. [6] utilize the finite element method for the theoretical analysis of multi-leaf springs. It was done to determine the safe load of the leaf spring and the fatigue cycles it can withstand. Alternate methods are utilized for the fatigue life analysis of leaf spring which comprises experimental fatigue life, Goodman's criteria, fatigue life based on analytical technique, and analysis using FEM. The experimentation is conducted on a 65si7 leaf spring material of a light commercial vehicle. Fatigue cycles to failure obtained from Goodman's criteria, analytical technique, and FEM are compared with the result obtained from the experimental analysis. The result shows that the fatigue cycle obtained from the analytical method and FEM deviates from the experimental method by 8.13% and 2.19% respectively. The research suggests that the graphical method is time-consuming and provides results in 8-9% variation. The CAE analysis results are closer to experimental results with an error of 2-5% of experimental results and can be used as an alternate method for a graphical approach which is reasonable. Hence, we are also considering the CAE analysis for the fatigue simulation of leaf spring. Apart from CAE analysis, very few analyses are done for the fatigue life analysis using the cumulative damage model. Baek et al [7], Hectors et al. [4], and Clarke et al. [1] have discussed the cumulative damage model using the Miners rule. As the leaf spring component undergoes a variable amplitude loading, [1,7,4] suggests the use of a Rainflow cycle counting algorithm to deal with Miner's rule under variable loading conditions. The study by Baek et al. [7] suggests the use of a probabilistic model of cumulative failure rather than a cumulative damage model in terms of accuracy, but the scope of our study is limited to the cumulative damage model as no such studies are done for the fatigue analysis of leaf spring.

This study focuses on the fatigue life analysis of leaf springs under general highway road conditions. The data is extracted from a public vehicle operating on these roads. Initially, the Rainflow cycle counting method and, Miner's rule is used for fatigue analysis. The results are then validated using the history data in CAE for fatigue analysis.

2. Design and fabrication

The casing of the data collecting module is 3D printed. Electrical components like batteries, switch, and control system components like ultrasonic sensors, and Arduino Uno are installed on the data collecting module. Arduino is programmed using C++. An ultrasonic sensor measures the deflection of the leaf spring, and the measured values are stored on an SD card with the help of the Arduino program.

Figure 1. Fabricated data collection module

3. Conceptual design and data collection

Data collection is done using a fabricated data collection module which consists of supporting units, power units, and control units as shown in Figure 2. Initially, the vertical distance between the main leaf spring and the ultrasonic sensor is recorded. After power is supplied to the module, it continuously measures the vertical distance between the main leaf spring and the ultrasonic sensor. The ultrasonic sensor measures the deflection of the leaf spring every ten times per second. The measured distance is recorded on the micro-SD card with the help of Arduino Uno. The recorded data is imported into the Excel sheet. For every data point, the initially measured distance is subtracted from it, which now gives the deflection of the leaf spring. The average is taken for the data points for every second, by which an average deflection of leaf spring is obtained for each second.

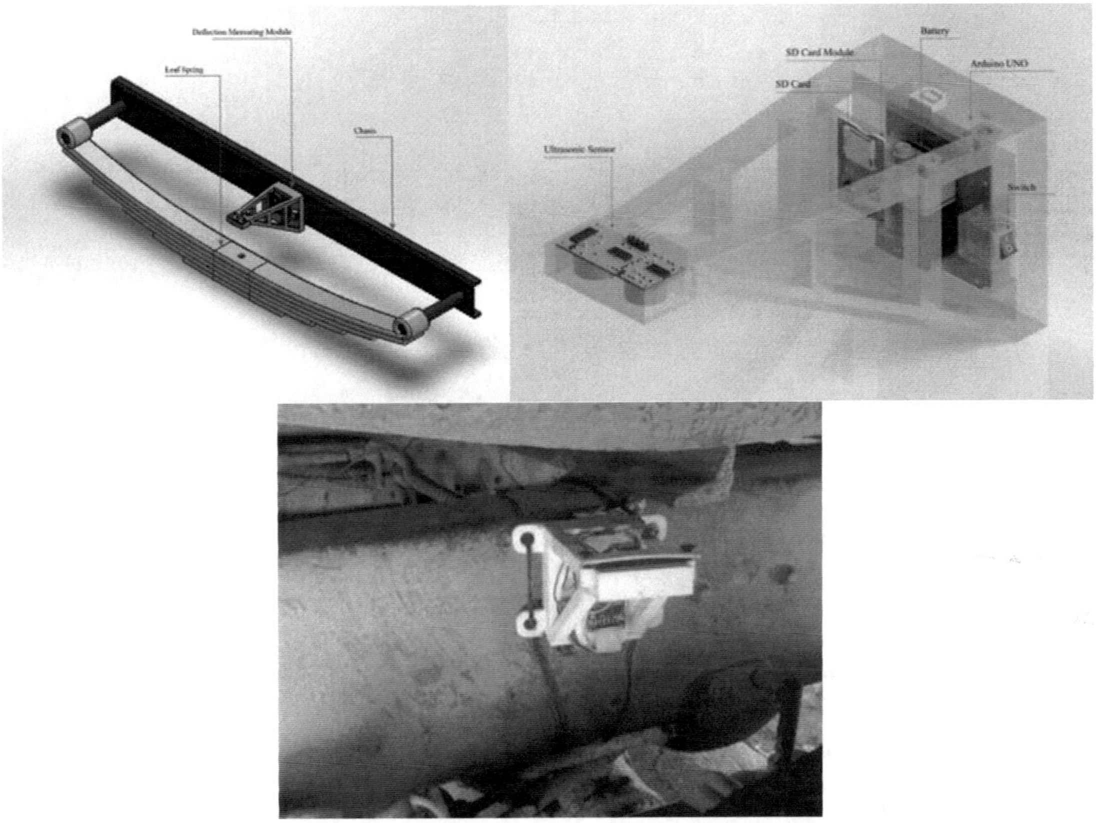

Figure 2. Conceptual design of data collection model and data collection

4. Methodology

There have not been any relevant studies regarding the fatigue life of a leaf spring in the context of Nepal. In Nepal, public vehicles operate mainly on paved roads and gravel roads under variable loading conditions. The road tested in this paper includes an urban road ranging 60km from Ratnapark to Sundarijal. The urban road selected for the study is representative of most of the roads found in Nepal. Urban roads consist of various irregularities such as varying degrees of slopes, potholes, and cracks which leads to variable road conditions. This is due to the poor maintenance of the road and heavy monsoon rain in Ratnapark. Thus, the proper standardization for the materials of leaf springs, according to the road conditions of Nepal is unknown. This paper aims to figure out the fatigue life of a leaf spring of a public vehicle (TATA-709) traveling in the local route of Sundarijal to Ratnapark.

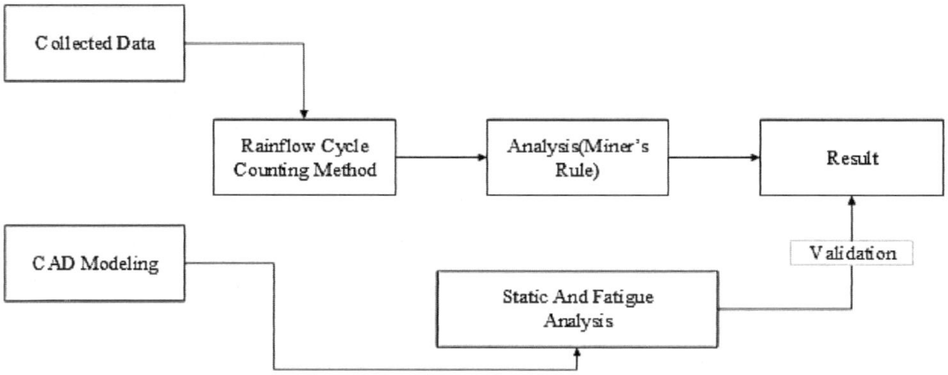

Figure 3. Methods for Fatigue Life Analysis

4.1. Fatigue life estimation by graphical method

Here, the study is done by graphical method or stress-life approach. In this method initially, the deflection data of the leaf spring is collected by the developed data collection module. The bending stress of the leaf spring is calculated by using the bending equation.

$$\frac{M}{I} = \frac{\sigma}{y} = \frac{E}{\rho} \quad (1)$$

$$\sigma = \frac{Ey}{\rho} \quad (2)$$

where M is the bending moment, I is the moment of inertia, σ is bending stress, E is Young's modulus of elasticity, y is a deflection from the resting point, and ρ is the radius of curvature of the leaf spring.

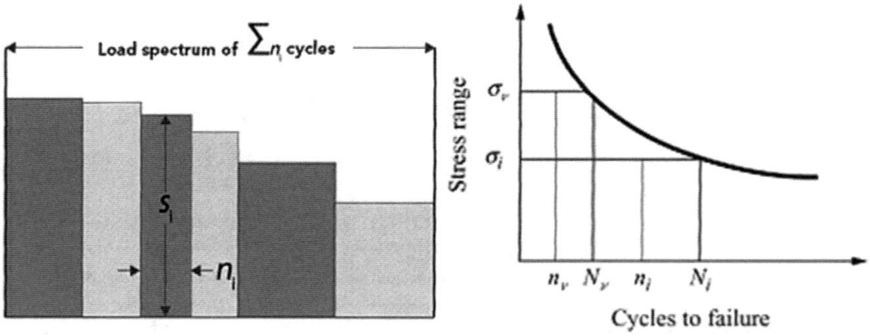

Figure 4. Load spectrum Vs stress range (left) S-N curve (right) [3]

The stress-time graph is obtained using the bending stress that was determined using the bending equation. There are a lot of fluctuating stresses and highly irregular data generated by the stress-time graph. To convert a loading sequence with varying stresses into an equivalent series of stress reversals with constant amplitude, the rain-flow counting approach is used. Here, the fatigue of the test specimen is investigated in the frequency and temporal domains. The time history is also used to calculate the number of cycles. Miner's rule analysis is used to get fatigue life. The Rainflow-counting method is used to convert a loading sequence with variable stress into an equivalent series of stress reversals with constant amplitude to calculate the fatigue life of a component. The fatigue of the test specimen is analyzed in the frequency and temporal domains [3]. Rainflow cycle counting is investigated in the time domain by using the load's temporal history as an input.

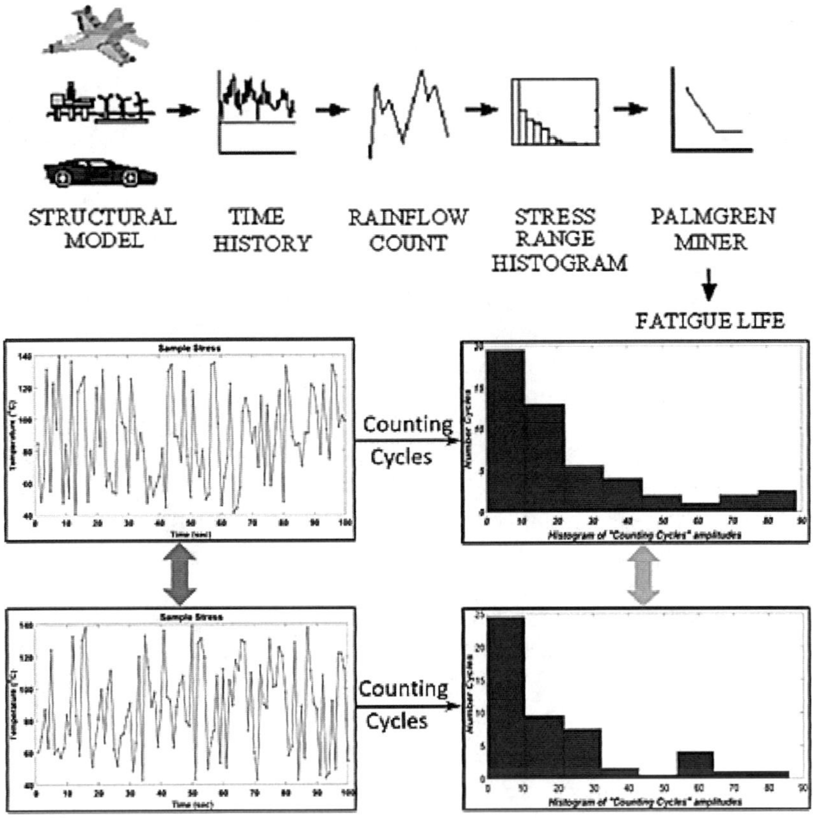

Figure 5. Rainflow cycle counting method [5, 12]

The number of cycles is computed using the temporal history. To break down the chaotic temporal histories into equivalent block loading stress, Rainflow cycle counting is used. A stress range histogram usually displays the number of cycles in every block. The fatigue life can be computed using the Miner's rule analysis [3]. The basis of Miner's rule, sometimes referred to as the linear damage rule, is the idea that fatigue damage is equal to the accumulated cycle ratio, which is shown in equation (3).

$$D = \sum_{i=1}^{n} \frac{n_i}{N_i} \tag{3}$$

n_i and N_i in equation (3) are the number of applied cycles and the number of cycles to failure for the i_{th} constant amplitude stress level, respectively, and D is the damage (where D equal to unity theoretically corresponds to failure) [10]. With varying fatigue cycles, Miner's rule determines the percentage of damage caused by varying stress levels. Failure is defined as occurring when the total of the damage fractions exceeds 1.0 [10]. From the graphs of stress versus time and stress versus number of cycles, respectively, n_i and N_i are obtained.

4.2. Fatigue life estimation by FEM in Ansys

The leaf spring's standard drawing is used to create the CAD model of each leaf, which is then assembled. To process the leaf spring assembly further, it is put into the analysis program ANSYS. The model meshes into little parts and the various contacts are configured. The solver provides the solution once the boundary conditions are applied while accounting for the real loading conditions [13]. Two analyses are carried out in ANSYS.

IOP Conf. Series: Materials Science and Engineering **1314** (2024) 012002 doi:10.1088/1757-899X/1314/1/012002

- Certain characteristics, including material attributes, loading circumstances, support conditions, and contacts, are considered in static structure analysis and must be provided as input for the analysis's pre-processing. The whole load is split between the two eyes of the master leaf when the CAE boundary conditions are applied while accounting for external loading conditions.
- Although leaf springs might function effectively at first, fatigue failure brought on by repeated cyclic stress frequently causes them to fail in service. The purpose of fatigue analysis is to characterize a material's ability to withstand the numerous cycles that a leaf spring may go through throughout its lifespan. The stress life and strain life methods are the two possible methods in the ANSYS fatigue module. The fatigue life of the component is predicted for this task using the stress life approach. By applying the boundary requirements and accounting for the actual loading conditions, the static environment is established. The fatigue tool is used to estimate fatigue life after static analysis.

5. Experimental Analysis

The experiment is conducted in a TATA-709 bus running in Ratnapark. Figure 6 shows the stress Vs time plot for the leaf spring during two trips of the vehicle (Sundarijal - Ratnapark) in the urban road conditions (City Road) of Ratnapark Nepal. Data was collected from a data-gathering module mounted on the leaf spring of the vehicle. The vehicle ran a total of 45 km in 56 mins. Two separate data were collected over this period; the total number of collected data sets was 46,000, each being deflection values of the leaf spring. High peaks and valleys are observed in the data which is very erratic due to variable amplitude loading. High peaks and deep valleys are observed mostly during braking and due to the physical conditions of the road.

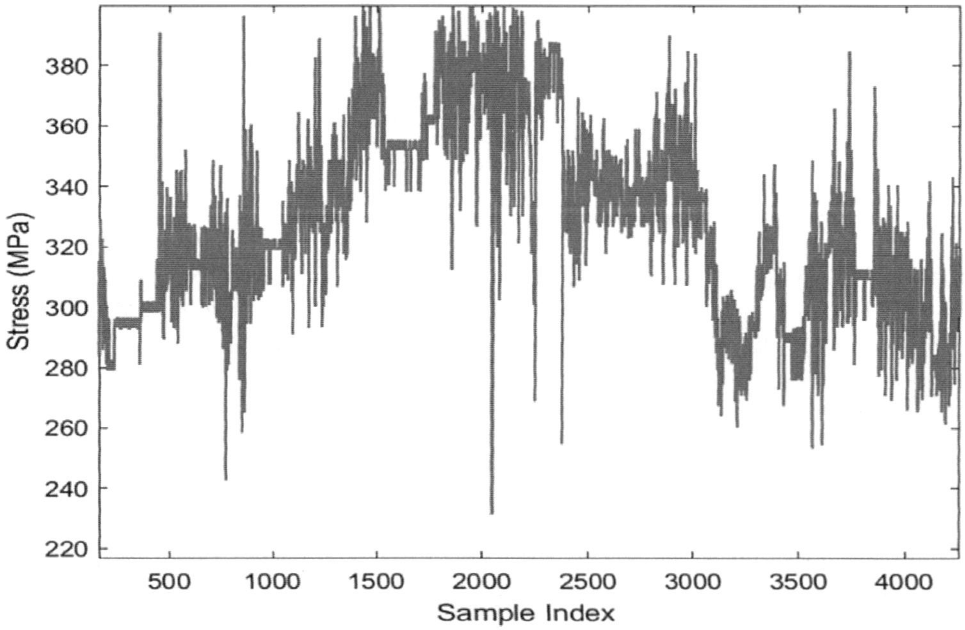

Figure 6. Stress (MPa) Vs sample index (sec)

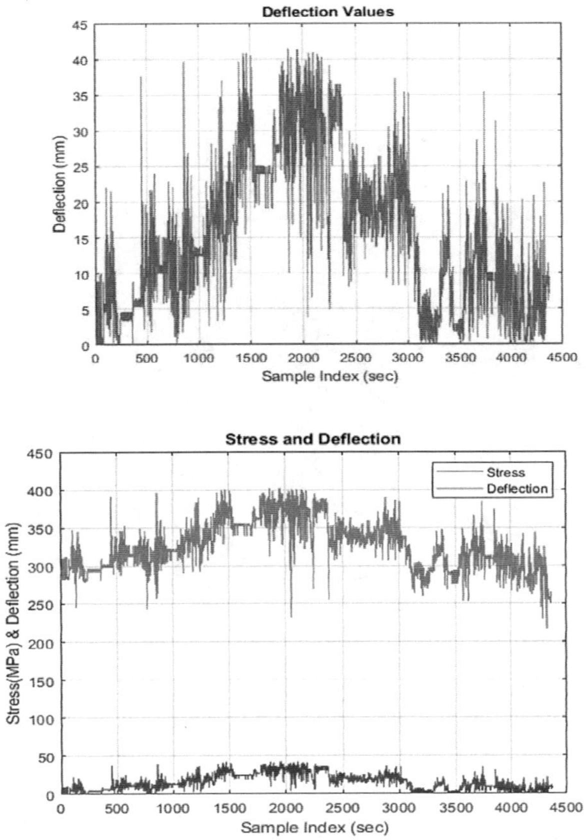

Figure 7. Deflection of leaf spring (top) stress and deflection (bottom)

Figure 7. displays the effect of variable amplitude stress on the deflection of leaf spring. During the trip cycle, the stress experienced by the leaf spring is highly erratic and highly varying due to changing road conditions. This varying stress results in significant deflection of the leaf spring in the Y direction. The combined analysis of the Stress vs. Sample Index curve and the Deflection vs. Stress curve shows that the leaf spring undergoes deflection in the Y direction when the vehicle encounters bumps and other irregularities. The leaf springs must absorb these high-amplitude shocks to maintain the vehicle's vibrational stability and overall performance. The continuous and varying stress leads to a consistent pattern of deflection in the Y direction, as the leaf springs work to mitigate the impact of the road conditions.

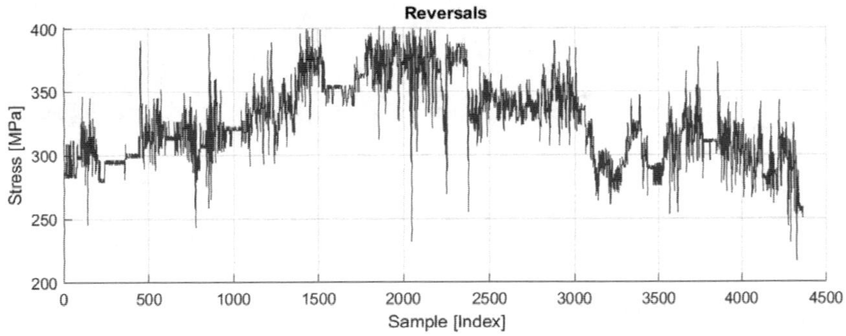

Figure 8. Stress reversals

Figure 8 displays stress reversals for variable stress time history displayed in Figure 6. It plays a huge role in implementing Rainflow cycle counting methods to generate stress range and frequency histograms. Extrema in the stress time curve are identified by implementing a MATLAB program. Which then replaces the waviness of the curve with a possible straight line. It connects two extreme points having the highest differences and ignores the value that lies between those points.

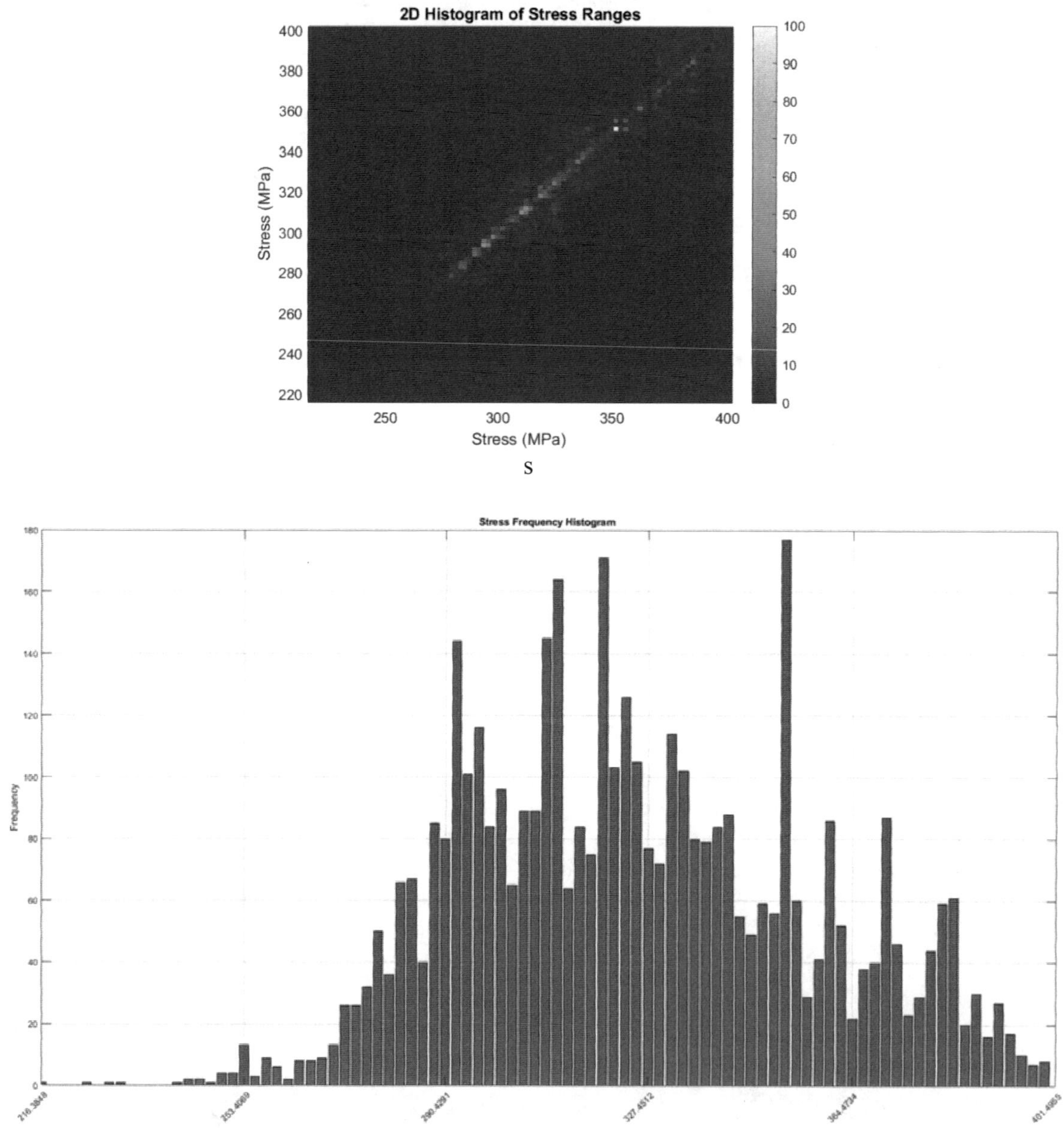

Figure 9. Rainflow matrix (top) stress range in MPa and 2D Rainflow histogram (bottom)

Figure 9 (top) resembles the 2D Rainflow matrix. A MATLAB in-built function based on ASTM (American Society for Testing and Materials) standards is implemented for the Rainflow cycle counting method for varying stress ranges. The program identifies the maximum and minimum values in the provided varying stress amplitude data and applies the Rainflow cycle counting method accordingly. The program then generates a square matrix for the given stress values. The plot helps to visualize the

IOP Conf. Series: Materials Science and Engineering **1314** (2024) 012002 doi:10.1088/1757-899X/1314/1/012002

distribution of stress ranges in the given data. It shows the relationship between each stress value and its adjacent stress value. The color intensity indicates how frequently different stress ranges occur. It shows that for the stress range of 300 – 400 MPa, the intensity of the stress cycle is very high. The stress cycles below 300 MPa and above 400 MPa are very low. Then the generated data from the 2D Rainflow matrix is used to plot the stress range and frequency histogram as shown in Figure 9 (bottom). With the help of Figure 9 (bottom), we can more easily visualize the stress range and its frequency during a cycle. Figure 10 displays a Rainflow cycle histogram which resembles stress range, mean stress, and no of cycles implementing a MATLAB program using the ASTM standard. We can observe that lower stress cycles occur at high frequency over the higher stress cycles. Although lower stress cycles occur at lower frequencies, they contribute to significant damage fraction. Similarly, the stress ranges where more frequency is concentrated display adverse damage fraction in the leaf spring element.

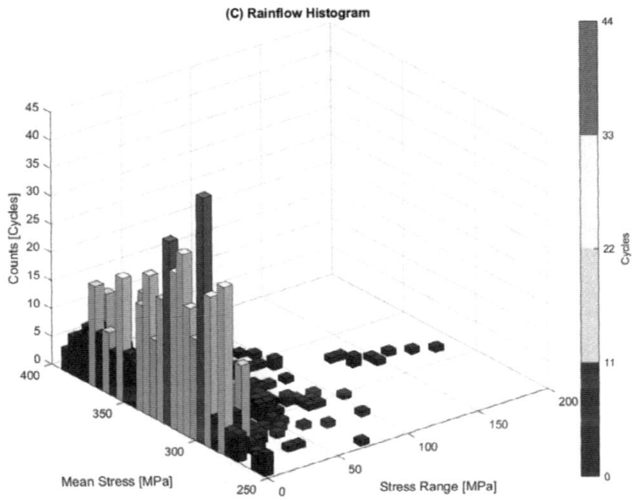

Figure 10. Rainflow histogram

Figure 11 corresponds to the stress and no of cycles curve technically known as the S-N curve. The fatigue S-N curve is drafted using the Rainflow cycle counting histogram in Figure 9. For a given stress range (or stress range vector) the program returns the fatigue endurance (or fatigue endurance vector) by implementing the computation based on ASTM standards. The result shows that the endurance limit for the leaf spring component under the provided environment is 71 MPa with 10^8 endurance cycles.

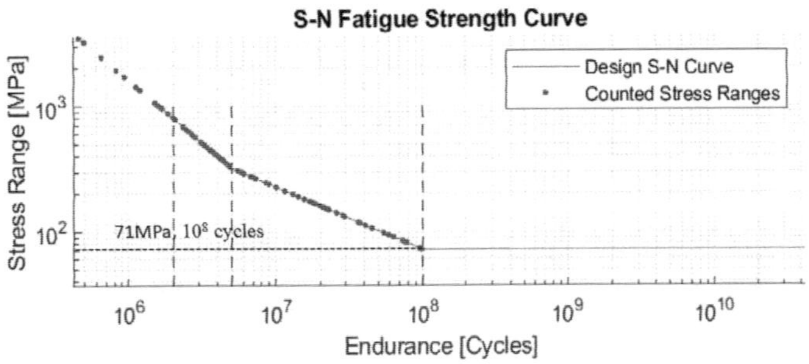

Figure 11. S-N curve

Figure 12 summarizes the accumulated fraction of damage in a leaf spring element based on the Palmgren-Miner linear damage accumulation rule. As discussed earlier, the stress time history recorded during the 56-minute trip was analyzed using the Rainflow counting method to extract stress cycles. These stress cycles are compared to the S-N curve for the material to determine the number of cycles that would fail at these stresses. The total damage fraction of 3.1091×10^{-5} was obtained using the Palmgren-Miner rule. This means that the leaf spring element has used 1.41×10^{-7} of its life during the trip and 9.9997×10^{-1} of its life remains. Therefore, the leaf spring will possibly undergo a fatigue failure after 8.046×10^5 stress cycles in the same route.

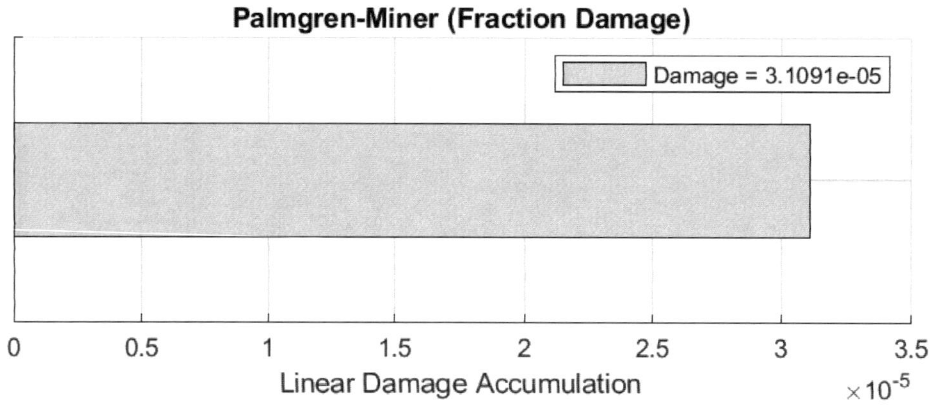

Figure 12. Linear damage accumulation

6. CAE analysis

The CAE (Computer Aided Engineering) analysis also validates the analytical or experimental simulation. Finite element analysis is a type of CAE analysis for structural analysis. FEA is a numerical method in design simulation where a complex system is broken into small elements and the reaction of each element to real-world forces, vibration, and other physical effects is observed. In this report, ANSYS is used to perform a Fatigue simulation of a leaf spring assembly of a vehicle. Different consecutive steps are followed during this process.

6.1. Material selection

The material 65Si7 is used for the fatigue simulation. Since the material is not in the ANSYS material library, the material property must be manually imported. The SN curve and other properties like density, tensile strength, and Young's modulus are adapted from Bhusan et al. [6] as illustrated in the table below.

Table 1. Mechanical properties of 65si7 material

Mechanical Properties	*Adopted Values*
Youngs Modulus	2.0012×10^5 *MPa*
Poissions Ratio	*0.266*
Bulk Modulus	1.4254×10^5 *MPa*
Shear Modulus	*79038 MPa*
Compressive Ultimate Strength	*1272 MPa*
Compressive Yield Strength	*1081.2 MPa*
Tensile Ultimate Strength	*1272 MPa*
Tensile Yield Strength	*1081.2 MPa*

6.2. Geometry and meshing

The dimensions for the CAD (Computer Aided Design) model are obtained by measuring the required parameter of an actual leaf spring used in a vehicle. The geometry is then inspected in ANSYS, and repair tools are used for any error in geometry. Meshing is the process of the process of dividing a complex geometry into a mesh of simpler shapes. The geometry of the leaf spring resembles a beam, sweep mesh is used for the simulation among Tetrahedron, Prism/ Wedge, and Pyramid due to its quadrilateral displacement. The mesh size of 8mm is used which generated 53813 elements with 231247 nodes. The mesh size is selected through an iterative process to make the simulation mesh independent as possible as shown in Figure 13.

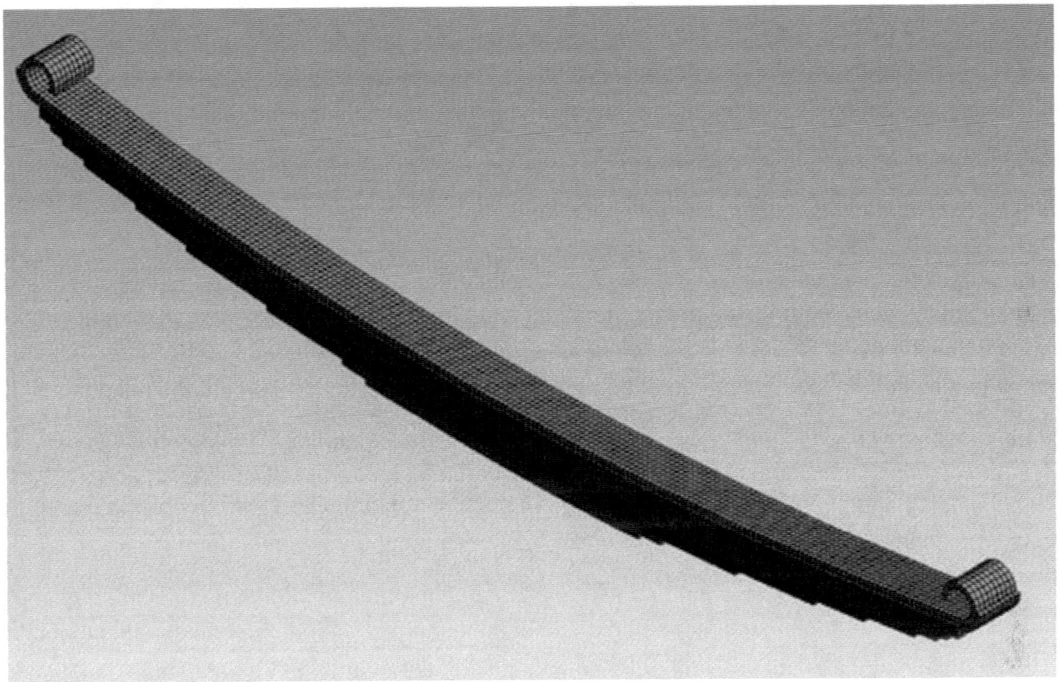

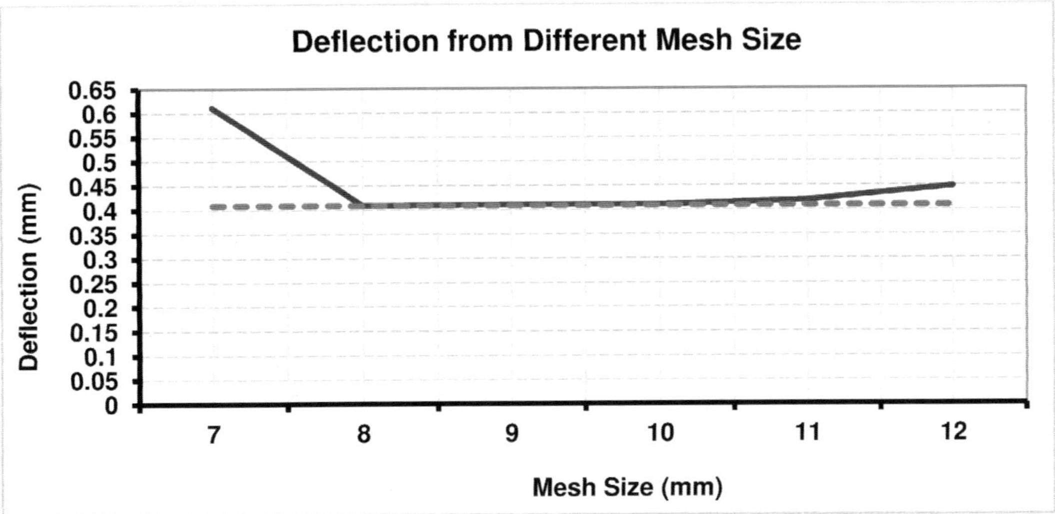

Figure 13. Meshing (top) mesh independence test(bottom)

6.3. Boundary conditions and static simulations

The boundary condition is set up to perform static simulation, which later creates the basis for fatigue simulation. Force is applied from the bottom in Y- the axis and fixed support is created eye of the leaf spring as shown in Figure 14. In Figure 14, B is fixed support and A is force.

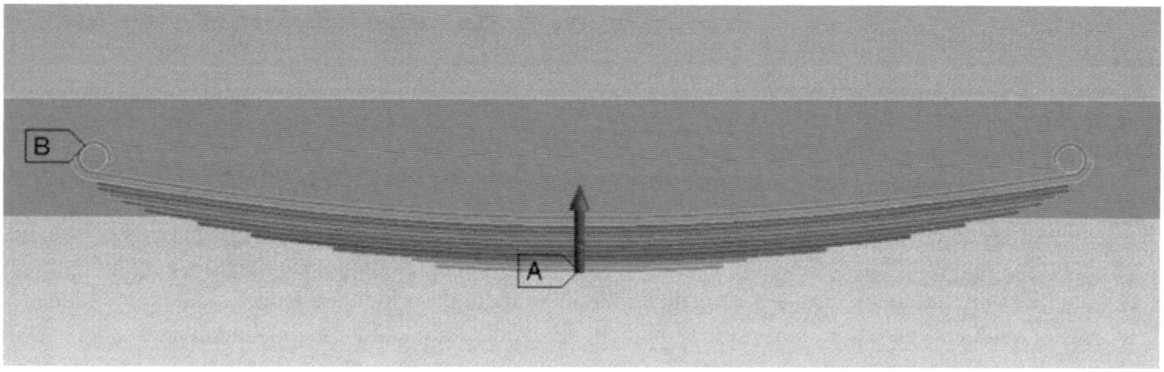

Figure 14. Boundary conditions

Static simulation was used to study stress, strain, deformation, and the validation of boundary conditions. This study was done by using static forces rather than dynamic forces. The function of certain static forces applied in these simulations has been derived from experimental deflection data. The simulated deflections of resulting forces were compared to the deflection data obtained experimentally, as provided in Figure 15. From this comparison, the deflection in simulation matched the experimental value very well, validating the statistical static force selected for further fatigue simulation. This agreement between the experimental and numerical results confirms that the deflection does not vary under static forces, thus giving a dependable basis for further analysis.

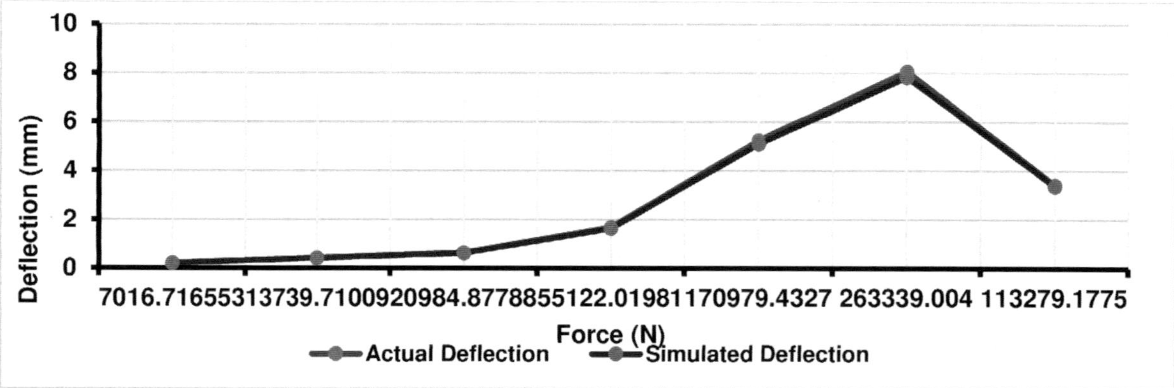

Figure 15. Comparative chart for actual and simulated deflection

6.4. Fatigue simulation

There are many parameters in fatigue simulation. The fatigue strength factor is set to 0.7 to compensate for geometry error which is taken from the literature for 65si7 material as mentioned by Bhusan et al. [6]. In this simulation, the calculated stress from experimental data is used as history data. For the simulation purpose, Goodman's theory is imposed for stress correction factor among Soderberg, Gerber, and ASME elliptical theory. Goodman's theory is generally used for the fatigue simulation of brittle material. Hence, we used the Goldberg theory for the fatigue simulation of 65si7 steel material of leaf spring.

IOP Conf. Series: Materials Science and Engineering **1314** (2024) 012002 doi:10.1088/1757-899X/1314/1/012002

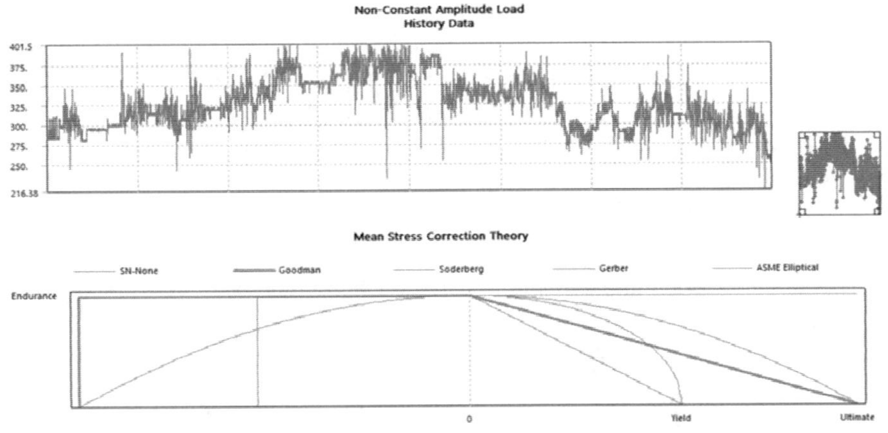

Figure 16. Fatigue simulation

The biaxiality tool in fatigue simulation is used to identify the stress type to proceed with fatigue simulation. Unity indicates biaxiality, zero indicates uniaxiality, and a negative one indicates pure share stress. Using the biaxiality tool, an appropriate stress condition is selected for fatigue simulation. The fatigue life and damage per cycle are then obtained from fatigue simulation.

7. Result and discussion

The deflection of the leaf spring is obtained for both the experimental method and CAE analysis in ANSYS. The leaf spring component undergoes a deformation in the Y direction due to the varying loading conditions. In our observations, there was no noticeable deflection in the X direction, as the primary deformation was observed in the Y direction, consistent with the bending behavior of the leaf spring under load. Figure 17 shows that the mid portion of the leaf spring element undergoes more deflection. The high deflection at the mid portion of the leaf spring is due to its maximum distance from the fixed support, which creates a greater bending moment and thus results in maximum deflection at this point. Therefore, the optimum distance from the fixed support to achieve maximum deflection is typically at the mid-span of the leaf spring.

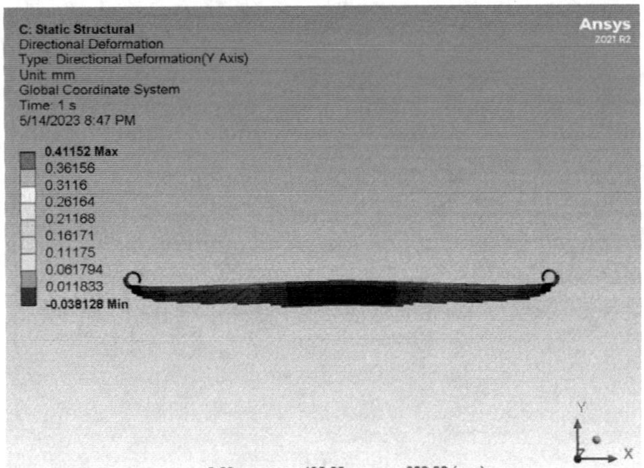

Figure 17. Deflection of leaf spring

IOP Conf. Series: Materials Science and Engineering **1314** (2024) 012002 doi:10.1088/1757-899X/1314/1/012002

Figure 18 shows the variation in the experimental data and the result obtained from the simulation for the deflection of the leaf spring. The deflection value is calculated for the different magnitudes of force experienced by the leaf spring during its cycle, for both experimental and CAE analysis methods. The result shows that the value of simulated deflection lacks by an average 2.8% margin. The result indicates that the value obtained for the deflection is very close for both the experimentation process. However, we have conducted a mesh independence study but still mesh refinement study can be done more accurately at the contact surface which requires high computing power.

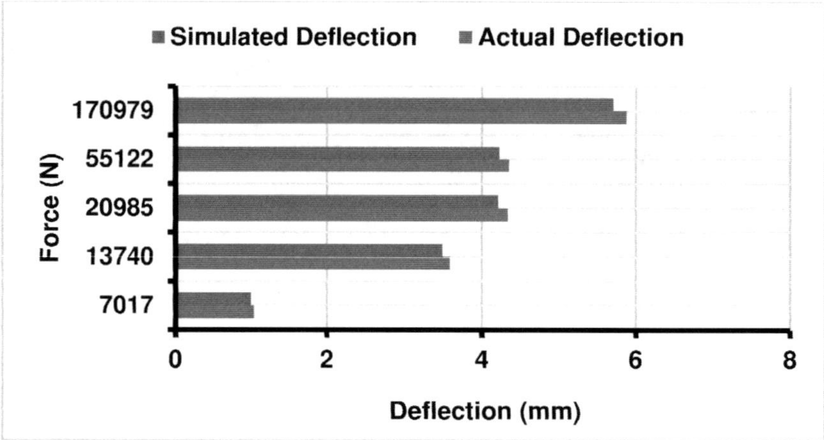

Figure 18. Comparison of deflection (in mm) for experiment and simulation for the same amount of force in (Newton)

In a practical scenario, the leaf spring is subjected to multiaxial stress. Biaxiality indication is the ratio of minimum principal stress to maximum principal stress. A biaxiality of 0 corresponds to uniaxial stress, -1 indicates pure shear and, 1 indicates pure biaxial stress. Figure 19 shows that during a one-cycle trip of a vehicle, the leaf spring element is dominated by the uniaxial stress.

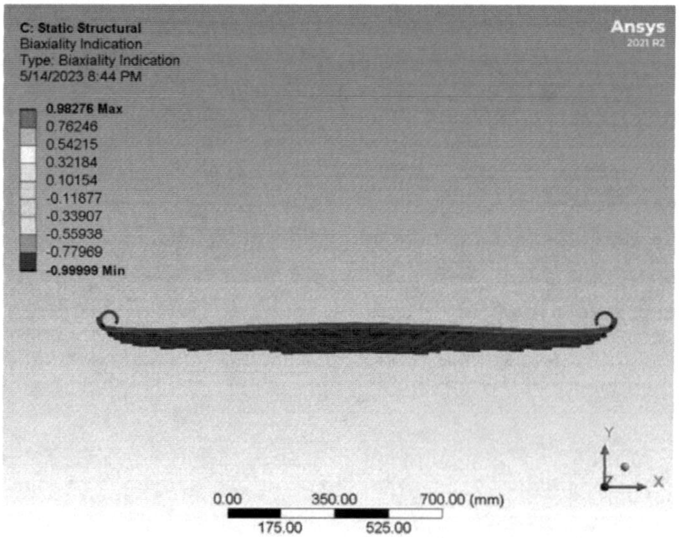

Figure 19. Biaxiality indicator

IOP Conf. Series: Materials Science and Engineering **1314** (2024) 012002 doi:10.1088/1757-899X/1314/1/012002

Figure 20 (top) shows the distribution of equivalent stress in the leaf at the prong area to failure. The maximum stress developed at the eye of the leaf spring. Hence the failure chances are more near the eye. Hence biaxiality condition is considered for that region as displayed in Figure 20 (bottom), which is in the uniaxial stress condition.

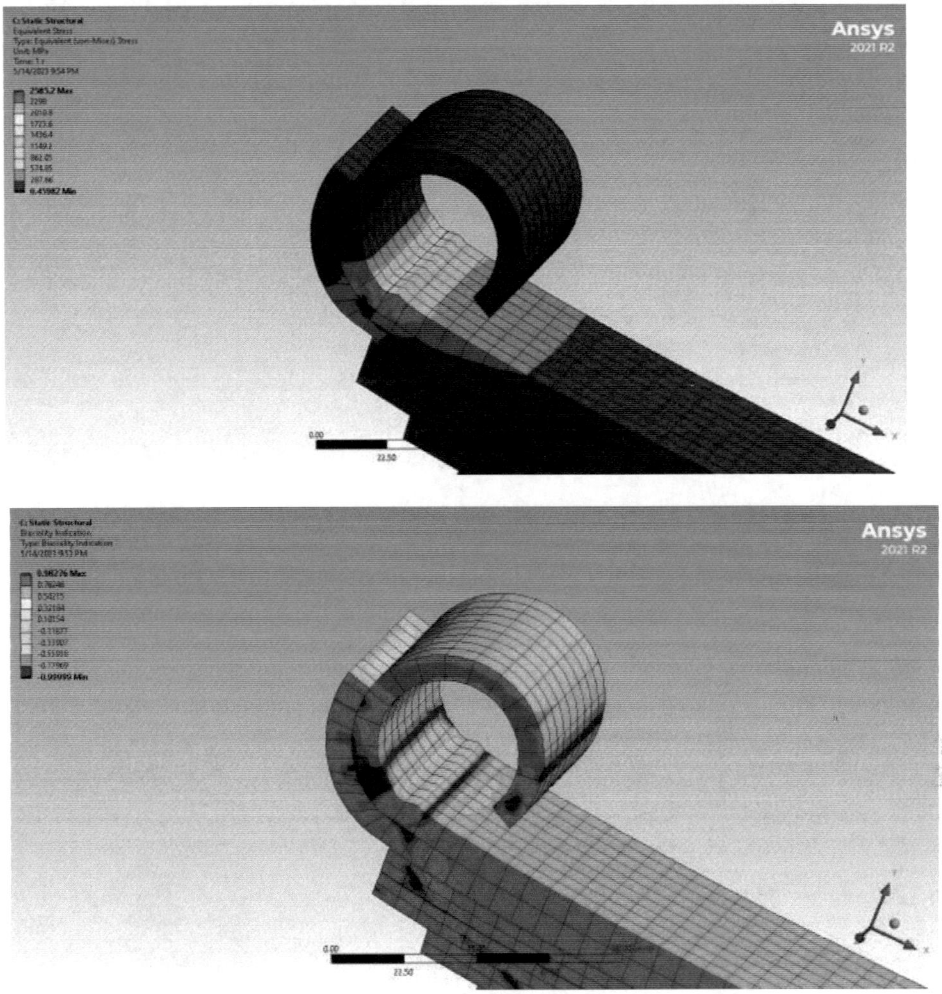

Figure 20. Equivalent stress at the pronged area to failure (top) biaxiality indicator (bottom)

Figure 20 (top) shows the distribution of equivalent stress in the leaf at the prong area to failure. The maximum stress is developed at the eye of the leaf spring. Hence the failure chances are more near the eye. Hence biaxiality condition is considered for that region as displayed in Figure 20 (bottom), which is in the uniaxial stress condition.

The time history data is fed into the fatigue analysis tool in Ansys software. After that, the number of cycles to failure is computed for the given stress history. The result resembles that for the provided stress history the leaf spring component can undergo 8.49×10^5 cycles before failure given the urban road stretching from Ratnapark to Sundarijal as shown in Figure 21.

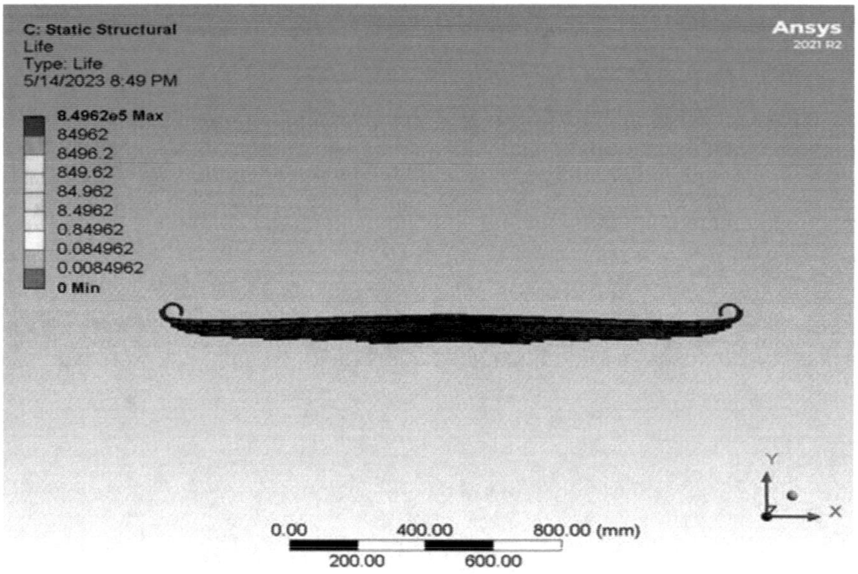

Figure 21. Life in a number of cycles

The study of Fatigue life obtained from Miner's rule corresponds to 8.046×105 cycles to failure. Whereas the fatigue life obtained from the simulated model of leaf spring is 8.49×105 cycles to failure as shown in Figure 21. A comparative study shows that the life of a leaf spring for the simulated model exceeds the life obtained from Miner's rule by 5%.

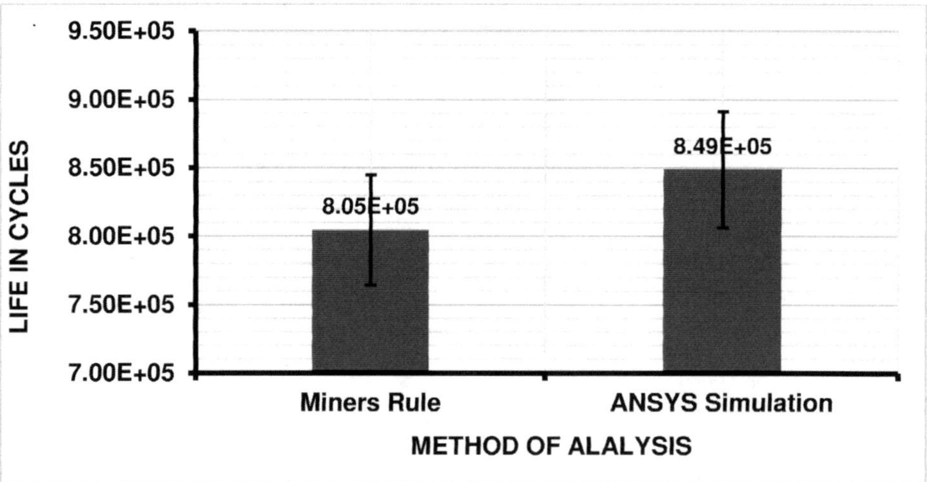

Figure 22. Comparison of life in cycles from the results obtained from Miner's rule and Ansys

The obtained life from Miner's rule descends by 5% from the obtained life from the simulation error in the ANSYS. The experimentation model is not conducted in the newly installed model of the leaf spring. It already has a certain amount of pre-stress and deflection value for the leaf spring element. Hence the life obtained from the Miner's rule is less. However, both models can determine the life of the spring with an allowable error range.

8. Conclusion

This study computes the number of cycles of failure of leaf spring element implementing Miner's rule and fatigue analysis in ANSYS. The result shows that the number of cycles to failure obtained from ANSYS exceeds the number of cycles of failure obtained from Miner's rule by 5%. The corresponding values for both methods are 8.49×10^5 cycles and 8.046×10^5 cycles. The obtained error suggests Miner rules underestimate the fatigue life compared to that of the FEA analysis performed in ANSYS. The discrepancy arises due to additional factors such as creep, rusting, pre-stress, and deflection already present in the leaf spring. These parameters are not considered in the study.

Although the conclusion of the paper points to the fact that this error could be due to pre-stress and deflection, which exist in the leaf spring, such pre-stress and deflection were not measured in this work. To be able to arrive at an accurate model, subsequent research on the newly installed models of the leaf spring should take pre-stress and deflection measurements. Also, data collection would be more reliable with the inclusion of RADAR sensors. Performing the following tests: creep, rusting, and effect of U-clamps could help in the analysis to arrive at further improvements in accuracy in predictions of fatigue life.

Acknowledgments
The authors would like to thank the Department of Mechanical Engineering, Kathmandu University for supporting the work. We would like to express our gratitude to Dr. Ranjit Shrestha and Mr. Uren Amatya for their constructive comments, we are also very much thankful to USC 2023 for providing us an opportunity to present this work.

References

[1] E. Giannakis, M. Malikoutsakis, and G. Savaidis, "Fatigue Design of Leaf Springs for New Generation Trucks," *IOP Conf. Ser. Mater. Sci. Eng.*, vol. 161, no. 1, 2016, doi: 10.1088/1757-899X/161/1/012065.
[2] M. Senthil Kumar and S. Vijayarangan, "Static analysis and fatigue life prediction of steel and composite leaf spring for light passenger vehicles," *J. Sci. Ind. Res. (India).*, vol. 66, no. 2, pp. 128–134, 2007.
[3] C. M. Sonsino, "Structural durability assessment of welded offshore K-nodes by different local design concepts," *Frat. ed Integrità Strutt.*, vol. 3, no. 9, pp. 3–12, 2009, doi: 10.3221/igf-esis.09.01.
[4] K. Hectors and W. De Waele, "Cumulative damage and life prediction models for high-cycle fatigue of metals: A review," *Metals (Basel).*, vol. 11, no. 2, pp. 1–32, 2021, doi: 10.3390/met11020204.
[5] S. Thillikkani and M. Nataraj, "Fatigue life prediction of heavy vehicle suspension system under varying load conditions," Advances in Mechanical Engineering, vol. 12, no. 11, p. 168781402096832, Nov. 2020, doi: https://doi.org/10.1177/1687814020968325.
[6] V. K. Arora, G. Bhushan, and M. L. Aggarwal, "Fatigue Life Assessment of 65Si7 Leaf Springs: A Comparative Study," *Int. Sch. Res. Not.*, vol. 2014, pp. 1–11, 2014, doi: 10.1155/2014/607272.
[7] C. K. Clarke and G. E. Borowski, "Evaluation of a Leaf Spring Failure Evaluation of a Leaf Spring Failure," vol. 5, no. December, pp. 54–63, 2005, doi: 10.1361/154770205X76303.
[8] A. Pendyala, C. Siva, R. Krishna, and C. Jagadish, "Fatigue life of composite leaf spring assembly Fatigue life of composite leaf spring assembly," 2021, doi: 10.1088/1757-899X/1185/1/012034.
[9] A. Dist, "STATIC ANALYSIS OF LEAF SPRING," vol. 4, no. 08, pp. 3794–3803, 2012.
[10] S. H. Baek, S. S. Cho, and W. S. Joo, "FATIGUE LIFE PREDICTION BASED ON THE RAINFLOW CYCLE COUNTING METHOD FOR THE END BEAM OF A FREIGHT CAR

BOGIE," vol. 9, no. 1, pp. 95–101, 2008, doi: 10.1007/s12239.
[11] R. Kuahal, A. S. Jayaram, and R. Chandan, "Design Analysis and Fatigue Life of a Leaf Spring Assembly," pp. 2393–2397, 2019.
[12] V. Samavatian, H. Iman-Eini, and Y. Avenas, "An efficient online time-temperature-dependent creep-fatigue rainflow counting algorithm," *Int. J. Fatigue*, vol. 116, no. April, pp. 284–292, 2018, doi: 10.1016/j.ijfatigue.2018.06.037.
[13] Y. S. Kong, M. Z. Omar, L. B. Chua, and S. Abdullah, "Fatigue life prediction of parabolic leaf spring under various road conditions," *Eng. Fail. Anal.*, vol. 46, pp. 92–103, 2014, doi: 10.1016/j.engfailanal.2014.07.020.

IOP Conf. Series: Materials Science and Engineering **1314** (2024) 012003 doi:10.1088/1757-899X/1314/1/012003

SIZING OF AN ON-SITE HYDROGEN REFUELING SYSTEM FOR A FLEET OF FUEL CELL BUSES IN DHULIKHEL- KATHMANDU ROUTE

Nawaraj Kafle[1], Tej Prasad Phuyal[2], Dhrubabar Singh K.C[2], Sagar Niroula[2], Nashla Sakhya[1] and Biraj Singh Thapa[1,2]
[1]Green Hydrogen Lab, Department of Mechanical Engineering, Kathmandu University
[2]Department of Mechanical Engineering, Kathmandu University, Nepal
E-mail: nawarajkafle1999@gmail.com, bst@ku.edu.np

ABSTRACT

Nepal's heavy-duty vehicle fleet relies entirely on imported fossil fuels, posing significant environmental risks due to high particulate matter (PM) emissions. Battery-powered alternatives are unsuitable for large fleets due to trade-offs between range and vehicle weight, which is not an issue for hydrogen-powered fuel cell electric vehicles (FCVs). Shifting to hydrogen fuel offers a promising solution for sustainable, emission-free heavy-duty vehicle operation. Fuel cell buses (FCBs) represent a transformative step in transportation, prioritizing environmental protection, decarbonization, and achieving net-zero carbon emissions by 2045. This paper reviews the critical components for developing an on-site hydrogen refueling station (OHRS). It analyzes the power requirements of vehicles operating on inclined roads based on vehicle modeling. By studying existing diesel buses on the route, the specifications needed for FCBs are determined. It involves Excel-based calculations of fuel demand, vehicle drive cycle modeling, infrastructure assessment, and capacity determination for a fleet of ten hydrogen buses. The analysis reveals that the maximum tractive power required is 119.85 kW, with buses averaging a velocity of 6.8 m/s, reaching a maximum of 20.75 m/s, and achieving a maximum acceleration of 0.96 m/s². The average motor power required is 13.09 kW, with a maximum of 151.48 kW. Each bus requires 8 kg of hydrogen per day, necessitating an OHRS capable of producing 80 kg per day for a fleet of 10 buses. Consequently, the sizing of components such as the electrolyzer, compressor, storage, and dispenser is dictated by a hydrogen production flow rate of 4 kg per hour. This study provides a comprehensive framework for implementing and operationalizing OHRS to support the deployment of FCBs in heavy-duty transportation.

Keywords: Onsite-refueling station, fossil fuel, vehicle modeling, Hydrogen-fuel, infrastructure assessment

1 INTRODUCTION

The rise in fossil fuel consumption and its negative impacts have become crucial topics today. Since 1970, CO_2 emissions have surged by about 90%, with fossil fuel burning and industrial activities responsible for about 78% of the total increase in greenhouse gases from 1970 to 2011[1]. The transport sector is a major contributor to these human-made emissions. In the fiscal year 2078/79, Nepal imported 1,723,557 KL of diesel, which is 2.3 times more than petrol imports and double the amount than 2077/78[2]. Emissions are a significant issue, with Nepal's total CO_2 emissions at 23 million metric tons in 2019, projected to rise to 34 million metric tons by 2030 and 79 million metric tons by 2050[3]. At COP26 in 2021, Nepal set an ambitious goal for net-zero carbon emissions by 2045 [4]. Achieving this goal will be tough without replacing diesel vehicles, as transportation remains a major challenge. Switching to electric and fuel-cell vehicles is essential. While petrol vehicles can easily be replaced by electric ones, diesel vehicles are the main source of emissions and fuel imports. The solution is to use fuel cell vehicles or hybrid vehicles that combine electric and fuel cell technology which need on-site hydrogen production and refueling station.

According to the AFC TCP data survey, there were 72,193 fuel cell vehicles worldwide by the end of 2023, with over 1,000 refueling stations available both on-site and off-site in their different development phase[5]. The 2023 update shows a significant increase in heavy-duty trucks, with

Asia, led by China, having 197 refueling stations. Currently, 80% of fuel cell vehicles are passenger cars, followed by buses,[6]. On-site hydrogen refueling stations produce hydrogen within the facility, requiring a production unit, hydrogen storage, and a dispensing unit. They primarily need an electricity source and water, ensuring that the hydrogen produced is zero-emission and classified as green hydrogen. Nepal has initiated studies on utilizing hydrogen from surplus electricity, solar, or other energy sources. On March 31, 2024, Kathmandu University commissioned a pilot-scale green hydrogen refueling station[7]. Despite hydrogen's potential as a clean and versatile energy source, several technical and economic challenges must be overcome to make hydrogen a viable option for widespread use. Additionally, significant investments in infrastructure are required to support the development of a hydrogen economy.

This report explores the potential of hydrogen fuel-cell vehicles as a sustainable solution for Nepal's transportation sector, evaluating its viability and potential for heavy-duty and longer-range transportation [8]. The technology for on-site hydrogen refueling stations is still emerging, making obtaining the necessary data challenging.

Hence, to determine the accurate size of the bus, the primary velocity and time data of a diesel bus operating on the selected route were recorded. The data of 15 days were collected and used to develop an Excel-based vehicle model. The model calculates the total power required, maximum and minimum power consumption, motor power, and total energy consumed by the diesel bus. The power consumption data was then converted to the necessary amount of hydrogen. The vehicle modeling was conducted for one bus and the refueling station was designed to support 10 fuel cell buses. The Electrolyzer was sized based on the hourly production rate, and the compressor's power requirement was defined for the specified flow rate. Additionally, storage capacity and volume were calculated, and a suitable standard dispenser was selected. This report details the vehicle modeling process and on-site hydrogen refueling station sizes.

2 STATE OF RESEARCH

The development of hydrogen as a fuel began with Sir William Robert Grove, who invented the Grove Cell in 1839, marking the first practical battery generating continuous electricity through a platinum electrode in nitric acid [9]. English engineer Francis T. Bacon expanded on this concept, perfecting the Bacon Cell in 1959, which became foundational for NASA's fuel cells in rockets. NASA's investment since the 1960s led to significant advancements, producing smaller, more efficient hydrogen fuel cells for various vehicles. The 21st century saw notable milestones, including a hydrogen-powered submarine by the German navy in 2004 and the world's first hydrogen-powered train tested by China in 2015. Today, hydrogen cars like the Toyota Mirai and Hyundai Nexo are commercially available [10]. The first commercial hydrogen refueling station was opened by First Element Fuels to fill trucks in California's Port of Oakland, [11], [12].

In 2023, Europe saw the establishment of 37 new hydrogen refueling stations, increasing the global total to 921 by the year's end under operation. Europe had 254 hydrogen stations in operation, with 92 percent of the new stations capable of refueling heavy commercial vehicles. Hydrogen infrastructure is mainly concentrated in East Asia, Central Europe, and the US coasts, with Germany leading Europe with 105 stations. Japan, South Korea, and China have the highest numbers globally, though data reliability for China is uncertain. North America showed limited growth, adding seven new stations, mostly in California. Overall, only 37 countries have hydrogen refueling stations, with the infrastructure being unevenly distributed[13]. Outside of East Asia,

Central Europe, and the US coasts, hydrogen refueling stations are sparse. In Asia, hydrogen refueling stations are well-established in China, Japan, and South Korea. Other countries, including Nepal, are conducting research and exploring the potential for on-site hydrogen refueling stations. Various universities in Nepal have initiated studies on hydrogen and its possibilities, highlighting the connection between hydrogen and hydropower in the country [Ref [14], [15], [16]]. Moreover, hydrogen is highly versatile, as it can be utilized not only for fueling but also for producing ammonia, urea fertilizer, industrial chemicals, and even vegetable ghee. Developing a hydrogen plant can therefore be beneficial for a wide range of applications.

3 VEHICLE MODELING

The 29.5 km long Kathmandu to Dhulikhel route was selected for the study and the data from the diesel-operated bus operating in this particular route was logged using the GPS sensor. The bus route encompasses both urban and suburban driving conditions and spans a distance of approximately 29.5 kilometers for a single trip.

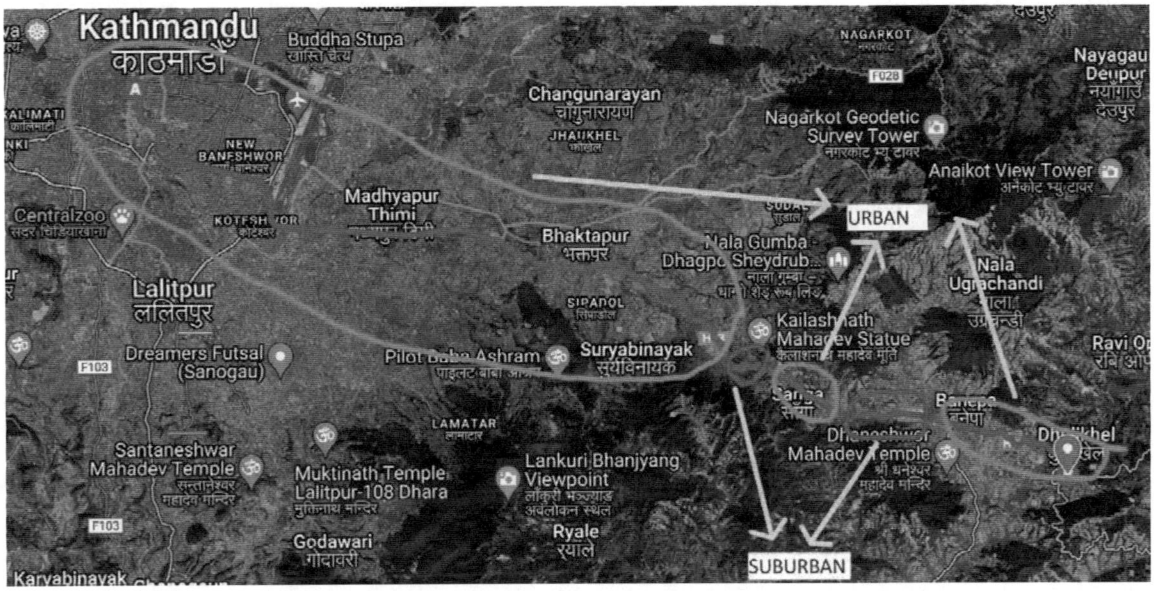

Figure 1: The selected bus route in Dhulikhel-Kathmandu route, Nepal

The round-trip public transportation route consists of 53 stops in the route. Mayur Yatayat's 15 days drive cycle data from a GPS tracker is analyzed to determine parameters like speed and system performance including speed as a function of time, providing valuable insights into the performance of the transportation system.

Vehicle modelling required many parameters to analyze the output motor power, and required energy to run the vehicle. For the analysis of the parameters for the model bus, some of the standard data were taken from the literature review and some of the parameters were calculated using their standard formula. References were from an LP 1112 model bus for fixing the gross weight of the bus as 12670 kg[17]. Acceleration due to gravity was taken as standard as 9.81 m/s^2. Coefficient of rolling resistance was 0.01 [18]. Coefficient of air drag standard value was taken 0.4[19]. Road slope was taken 20% as per standard conditions. Frontal area was calculated by subtracting area with clearance height from the ground level[20]. For 45 seated LP 1112, Length was found to be

9,100mm, Width of 2,570mm, and Height of 3,100mm. The frontal area is calculated using equation 1.

$$\text{Frontal Area} = \text{Width} * (\text{Height} - \text{Ground Clearance}) \tag{1}$$

- Air density is constant = 1.2754 under standard temperature[21].
- Resistance force was calculated as 1,242.927 from its main formula.

$$\text{Resistance force } (F_R) = mgC_R\cos\theta \tag{2}$$

Along with many kinds of forces acting on a bus, grade force is one of them. It is simply a gravitational force acting on a vehicle. This force may or may not be exactly perpendicular to the road surface depending upon the presence of grade. It can be calculated by multiplying weight of the vehicle and acceleration due to gravity.

$$\text{Grade force} = mg\sin\theta \tag{3}$$

With the slope value of zero, the force is obtained as zero.

Transmission efficiency was taken standard from the mentioned reference as 0.86[22]. Motor efficiency was calculated as 0.92 according to the mentioned reference as it was illustrated that most of the motors operate at 50-100% of the rated load and maximum efficiency is usually near 75% of the rated load[23]. So, calculated motor efficiency is in equation 4.

$$\text{Motor efficiency} = \frac{75}{100} \times 123 \tag{4}$$

= 92.25%

Maximum power was also taken as reference from LP 1112 model bus as 123 HP [24]. Fuel cell efficiency was 60% [25]. The energy efficiency of a fuel cell is generally between 40 and 60%; however, if waste heat is captured in a cogeneration scheme, efficiencies of up to 85% can be obtained. Design speed was also taken from the LP 1112 model bus as 107 km/h [24]. Tyre radius was calculated 0.3985m using the Equation 6 from reference [26]. The section height is calculated using Equation 5.

- Rim diameter = 17.5inch, Rim width=235mm, Aspect ratio=75

$$\text{Aspect Ratio} = \frac{\text{Section Height}}{\text{Section Width}} \tag{5}$$

- Section height=176.25mm

$$\text{Tyre diameter} = (2 * \text{Section Height}) + \text{Rim Diameter} \tag{6}$$

- Tyre radius= 0.3985m

Given equations are used to analyze vehicle specifications. The results of this data analysis and vehicle modeling are presented in various graphs. These graphs show the required power, total energy consumption, driving cycle, and the maximum and minimum power utilized in relation to road gradient. The graphs are derived from modeling analysis of 15 days of GPS data from a vehicle traveling the route between Kathmandu and Dhulikhel, making two trips per day.

3.1 Velocity and Acceleration vs Time

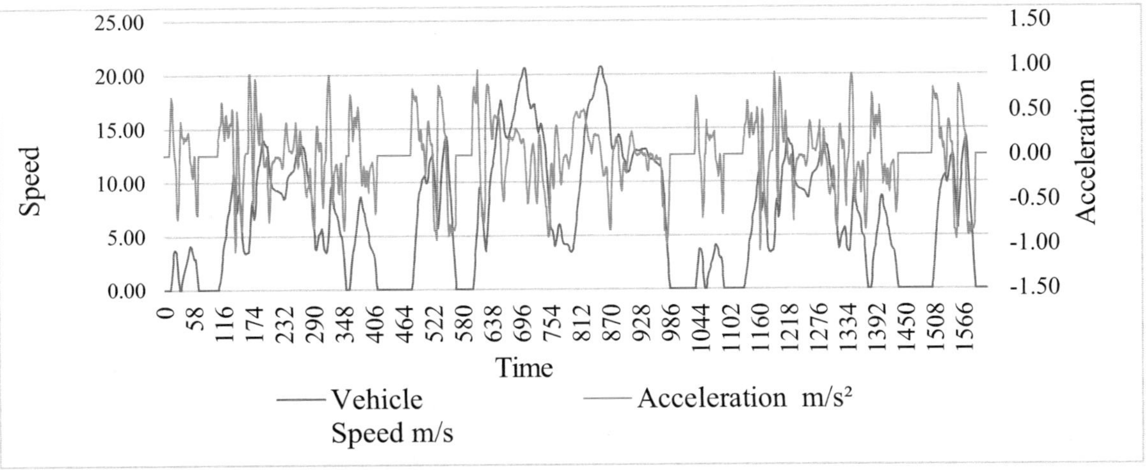

Figure 2: Graph of velocity and acceleration with time

Figure 2 shows the overall vehicle speed and its acceleration along the route over a given period. From the graph, it is observed that the maximum speed reaches 20.75 m/s, and the average speed is calculated to be 6.8 m/s, equivalent to 24.48 km/h. The graph also indicates a maximum acceleration of 0.96 m/s², suggesting that the road conditions in Nepal are relatively good. The vehicle's maximum speed was found to be 74 km/h on flat and wide areas but decreased sharply on inclined roads.

3.2 Tractive Power vs Time

Tractive force is the amount of total traction which is parallel to the direction of motion or can be stated as total pounds of pull that can be exerted before slippage occurs. It can be calculated by multiplying weight of the vehicle by the coefficient of traction.

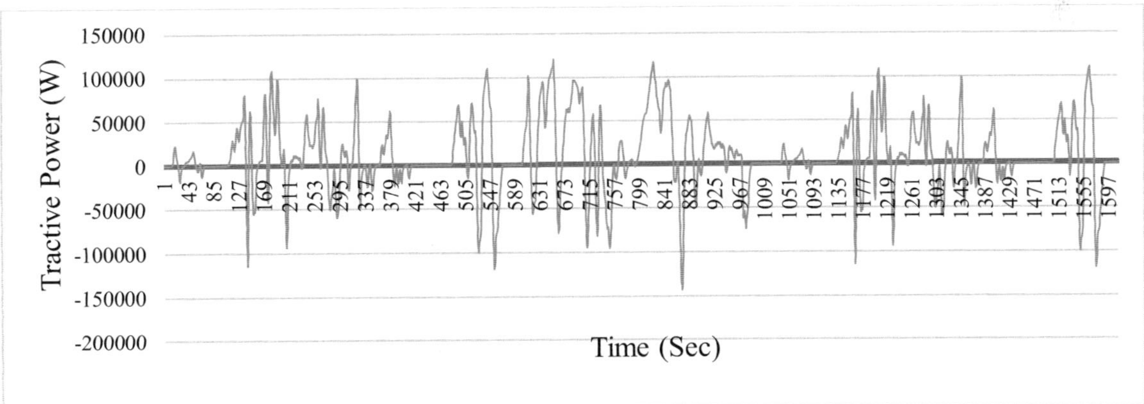

Figure 3: Graph representing tractive power to time

Tractive power is the total power supplied to the vehicle to overcome all opposing forces, such as drag force, friction force, and gravitational force. Figure 3 illustrates the required power under various road conditions—plain, inclined, and banked—over time. The average power required is

1,036.44 W; however, the maximum power 119,855.61 W is critical as the vehicle must deliver this power to navigate the road conditions successfully. Otherwise, the system will fail.

3.3 Total Energy Consumption by Motor

As the vehicle travels, it consumes energy based on the power needed for the route. Figure 4 displays the total energy required by the motor for the route, with the cumulative energy plotted over time. The graph shows that as time progresses and the vehicle covers more distance, its energy consumption increases accordingly.

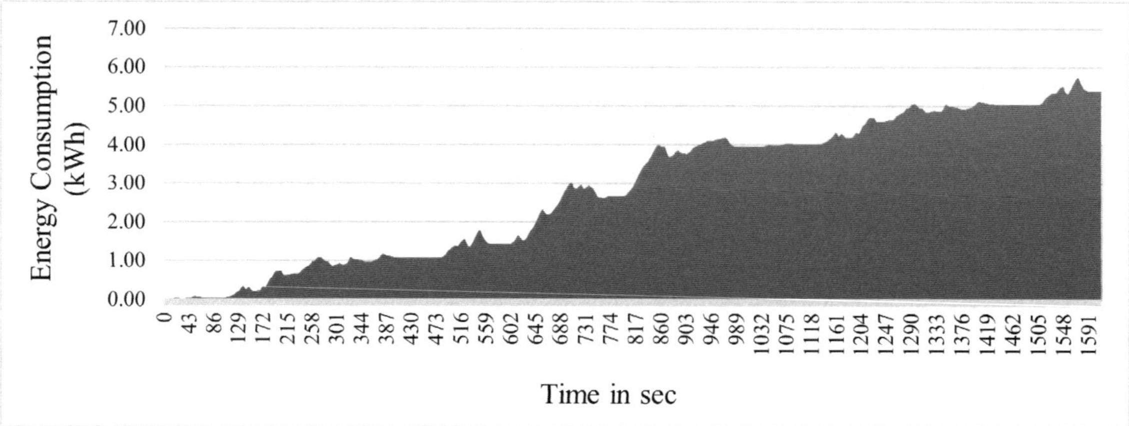

Figure 4: Graph representing total energy consumed by the motor with time.

Figure 4 visualizes the motor's total energy consumption in kWh, plotted against various time intervals in seconds. The graph indicates a maximum energy consumption of 5.4 kWh. Additionally, the total energy required by the vehicle for the route is calculated to be 59.0311 kWh.

3.4 Frequency Distribution Chart

The frequency distribution chart displays the percentage of motor operation within different power ranges. This data is useful for determining when the battery alone can be used, when the fuel cell should operate, and when both systems need to work together. In higher power ranges, both the battery and the fuel cell must operate to achieve maximum power and ensure the fuel cell runs within its most efficient range.

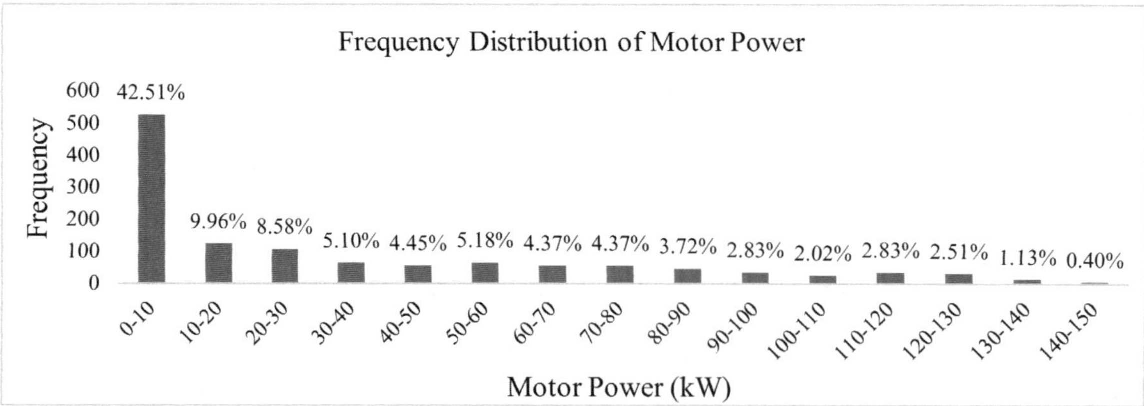

Figure 5: Graph representing the frequency of motor run on that power range

Figure 5 presents a frequency distribution graph of motor power relative to frequency, highlighting the speed ranges in which the vehicle operates most, least, and on average. This chart is essential for determining the vehicle's maximum power output, which in turn informs the necessary fuel cell and battery capacities. Over 50% of the motor's operation occurs under 20 kW, with a maximum power output of 151.48 kW. This information is crucial when selecting a hybrid bus. In the 0-10 kW range, about 42% of the operation frequency is observed, and as power increases, the frequency decreases, with the motor operating at 150 kW for only 0.4% of the time.

3.5 Frequency Distribution of Total Energy Consumption

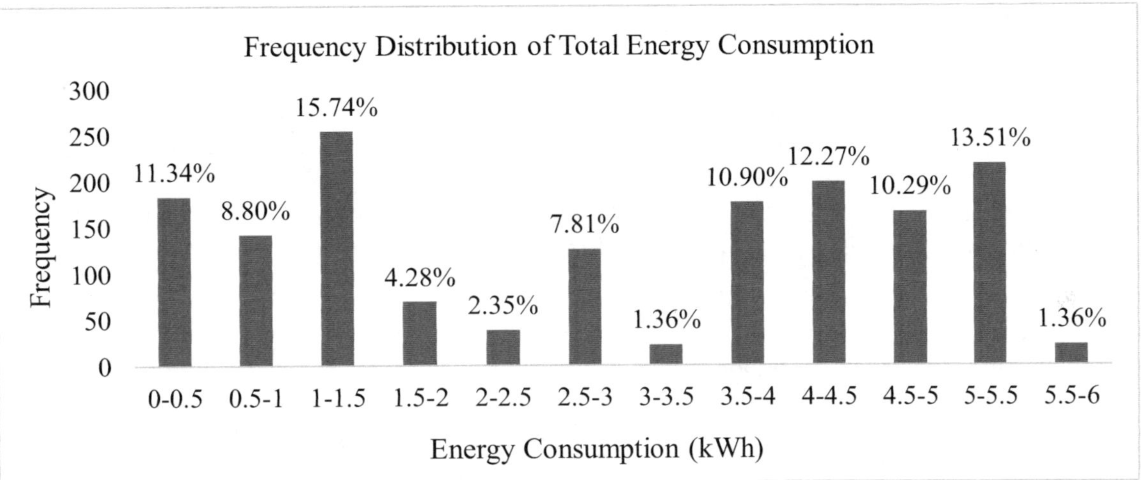

Figure 6: Graph showing energy consumption as a percentage over that range

Figure 6 depicts a frequency distribution graph of the vehicle's total energy consumption relative to frequency. This graph offers a snapshot of the percentage of total energy consumed at various power levels over specific time intervals. The highest frequency of energy consumption occurs in the 1-1.5 kWh range, followed by the 5-5.5 kWh range.

3.6 Obtained Result

Based on the calculated values and graphs, the average energy consumption of the vehicle was determined to be 0.4919 kWh/km when traveling a distance of 120km. Furthermore, the maximum energy requirement was calculated to be 59.0311 kWh, which will be used as a reference to find a bus with an appropriate power capacity within this range.
Motor power input average =13098.38W
Motor power output max= 151485.86W
Total maximum Energy Consumption =5.4kWh
Maximum average torque=2031.58 Nm
Maximum Torque =4415.76Nm
Based on the result the specific vehicle is selected which is capable to run in the selected route.
The specifications are in Table 1.

Table 1: Specification of LPO 1625 Hydrogen fuel cell bus [27]

Bus	Hydrogen Fuel Cell Bus
Fuel Cell Power System	Hydrogen Fuel cell

Fuel Cell Gross Peak Power	120 kW
Total hydrogen storage	29 kg
Peak Power Output of Motor	250 HP
Rated Torque	1050 Nm at 800 rpm
Number of cylinders	8 * 205 liter
Max. Speed	70 kmph
Gradeability	17%
Body Dimensions	12000 mm x 2600 mm x 3500 mm
Floor Height	390 mm

3.7 Total Requirement of Fuel

For the selected bus, with full tanks holding 29 kg of hydrogen, the vehicle can travel 450 km. This means 1 kg of hydrogen provides a range of 15.52 km. To cover 120 km in a day, 8 kg of hydrogen is needed. Therefore, 8 kg of hydrogen is sufficient for a bus to travel 120 km daily. With 10 buses in operation, a total of 80 kg of hydrogen is required each day.

4 PROCESS AND PLANT DESIGN

The plant sources electricity from the grid, which is then supplied to the electrolyzer to produce hydrogen. This hydrogen is temporarily stored in a buffer tank before being compressed to 350-400 bar by a compressor. The compressed hydrogen is stored in a tank and then pumped to the dispenser via a pump. A heat exchanger is used to manage the temperature of the hydrogen as it is produced, stored, and dispensed to vehicles. Figure 7 illustrates the complete value chain of the OHPRS, depicting the flow of electricity, water, and hydrogen.

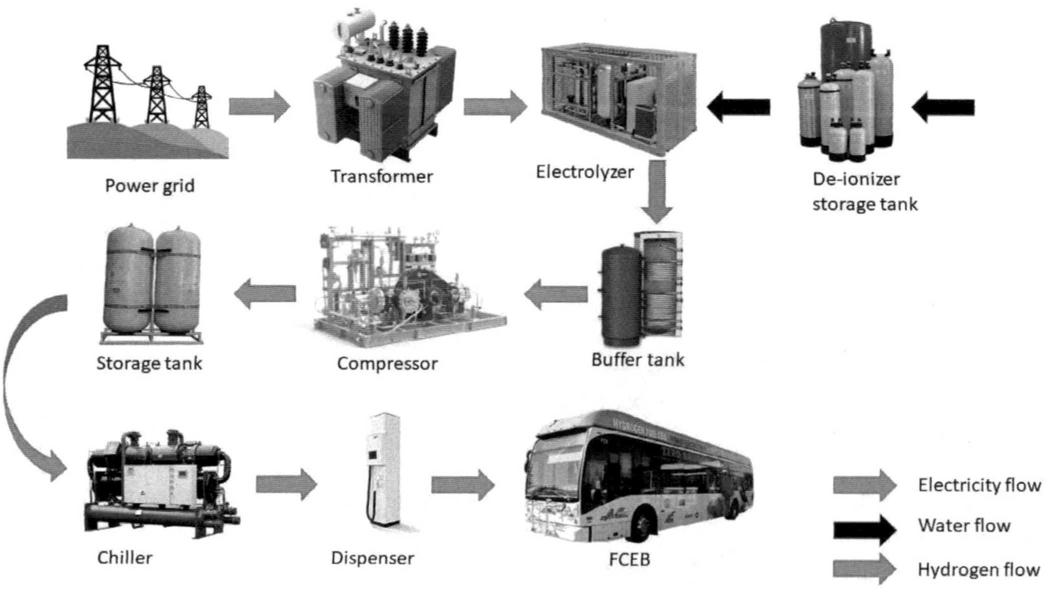

Figure 7: Process flow diagram of the overall system

Here is a step-by-step outline of the process and plant design for a hydrogen refueling station, including the anticipated P&ID (Piping and Instrumentation Diagram) for both the hydrogen production and refueling sections.

4.1 Anticipated PID of Hydrogen Production Unit

The anticipated Process and Instrumentation Diagram (PID) of a hydrogen production unit provides a comprehensive overview of the essential components and processes required to generate hydrogen. These include the transformer, rectifier, electrolyzer, water tank, gas separator, and other components, as depicted in Figure 8.

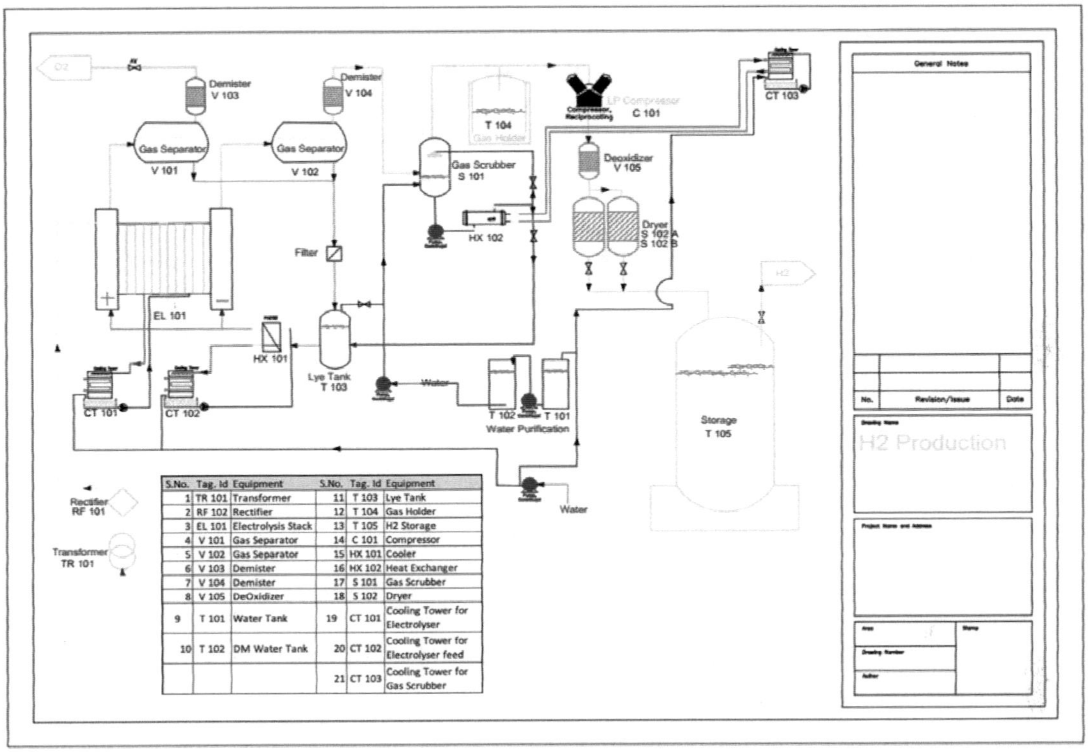

S.No.	Tag. Id	Equipment	S.No.	Tag. Id	Equipment
1	TR 101	Transformer	11	T 103	Lye Tank
2	RF 102	Rectifier	12	T 104	Gas Holder
3	EL 101	Electrolysis Stack	13	T 105	H2 Storage
4	V 101	Gas Separator	14	C 101	Compressor
5	V 102	Gas Separator	15	HX 101	Cooler
6	V 103	Demister	16	HX 102	Heat Exchanger
7	V 104	Demister	17	S 101	Gas Scrubber
8	V 105	DeOxidizer	18	S 102	Dryer
9	T 101	Water Tank	19	CT 101	Cooling Tower for Electrolyser
10	T 102	DM Water Tank	20	CT 102	Cooling Tower for Electrolyser feed
			21	CT 103	Cooling Tower for Gas Scrubber

Figure 8: Anticipated PID of hydrogen production unit

The anticipated PID values for a hydrogen refueling system on the system design and requirements. Hydrogen refueling systems aim for a pressure range of 350 to 700 bar, a moderate temperature range, and a flow rate that allows for efficient and timely refueling is presented in Figure 9.

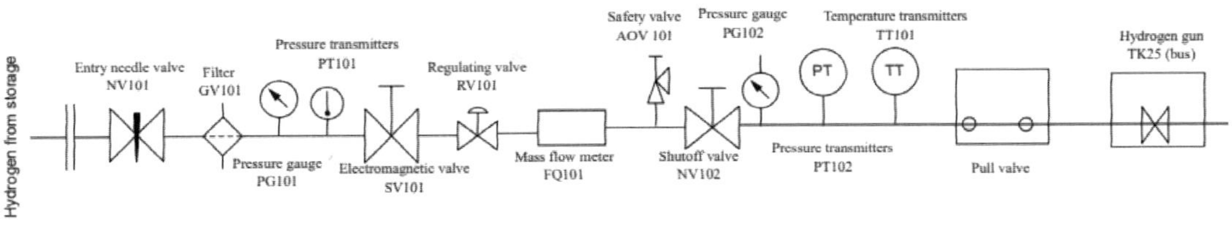

Figure 9: PID of Hydrogen refueling system

4.2 Components Selection

On-site hydrogen refueling station majorly contains the production unit, Storage unit and dispensing unit. Major components for production, storage and dispense are electrolyzer, compressed hydrogen tank and dispenser. As the production of hydrogen is 4 kg per day, the total electricity required is 250 kWh, the detail is in Table 2. Production units need a water treatment system, heat exchanger and electrolyzer. Different types of electrolyzer for green hydrogen production are, Proton Exchange Membranes (PEM), Alkaline Water Electrolyzer (AWE), Anion Exchange Membranes (AEM) and Solid Oxide Electrolyzer Cells (SOEC). Among, PEM and AWE are more commercialized and mature technology. PEM is more efficient, has less start time, has high purity of hydrogen produced, and has the capability to work on full load as well as half load with the same efficiency. So, the chosen electrolyzer is PEM over AWE. Storage of hydrogen can be done in different states and forms, widely used storage technology for vehicles is compressed hydrogen gas, for distribution, it is liquid hydrogen, for large amounts of hydrogen storage underground, salt caves are used. Hydrogen can also be stored in a solid state as reacted with different metals which is known as metal hydride, the gravimetric density is very low in this form but can be stored for a longer duration of time as dormancy time is very high. For small amount of hydrogen storage in less than 400 bars pressure compressed hydrogen tank are more suitable as they don't need more maintenance and can be dehydrogenated when needed easily. So, for 10 to 15 kg of hydrogen storage compressed hydrogen storage tank is best and if the availability of space is enough the storage can be done in type 1 or 2 hydrogen storage.

Table 2: Component selection

Components	Description	Type	Sizing	References
Water treatment system	Used for water demineralization, deionized, and other water purification processes	Mixed bed deionizer	Purification capacity 800 liter/day	[28],[29], [30]
Electrolyzer	Hydrogen production unit, Produce green hydrogen with the help of water and electricity. The only by-products are oxygen and unreacted water vapor.	Proton Exchange Membrane (PEM)	225 kW, 4 kg/h, 70% efficient, High part load and full load efficiency, good startup time	[31],[32], [33]
Compressor	Hydrogen compressor that compresses up to 500 bar pressures in multi-stage.	Multi stage diaphragm compressor,	85% efficient, 3 kW power	[34],[35], [36] [37]
Storage	Hydrogen storage medium is the most crucial part of OHPRS. It needs regular inspection.	Type two compressed hydrogen storage	The pressure of store hydrogen up to 350 bar, density 23.36 kg/m^3, the volume required 3.44 m^3.	[38], [39], [40], [41]

Dispenser	Use to dispense hydrogen at different pressures as required to the vehicle	Type 2, with optional Type 1 and other regional nozzle types	13 kW, maximum pressure up to 700 bar can be dispensed, 4-meter standard hose.	[42], [43]

5 CONCLUSION

The hydrogen refueling station (HRS) is essential for fuel cell vehicles, analogous to petrol stations for gasoline and diesel vehicles, and charging stations for electric vehicles. The vehicle modeling performed in this study indicates that the average energy consumption of the vehicle is 0.4919 kWh/km for a distance of 120 km, with a total energy requirement of 59.0311 kWh. The average motor power input was 13.09 kW, while the maximum motor power output was 151.48 kW. The maximum energy consumption is 5.4 kWh, with average torque of 2,031.58 Nm and maximum torque at 4,415.76 Nm. The maximum tractive power required for the bus is 119.85 kW, crucial for determining the power of fuel cell buses. The selected HRS caters to 10 fuel cell buses, producing 80 kg of hydrogen daily and refueling each vehicle with 8 kg of hydrogen. Key components of the HRS include the source of water and electricity for green hydrogen production, requiring 250 kWh of electricity and 720 liters of pure water. The production unit, a 225 kW PEM electrolyzer, generates hydrogen at a rate of 4 kg per hour. The storage unit, a type 2 compressed gas hydrogen storage, stores the hydrogen at 350 bars pressure with a volume of 3,430 liters. The HRS features a 4-meter-long standard hose dispenser with two nozzles, enabling it to refuel two buses simultaneously in a very short period.

An on-site Hydrogen Refueling Plant is crucial for a lower-income country like Nepal, which boasts abundant hydropower potential and is poised to become a net electricity exporter due to the rapid expansion of hydropower plants. The potential surplus of electricity can be effectively utilized by increasing demand through the establishment of hydrogen plants, which can then be used in the transportation sector. For example, deploying just 10 buses on a specific route could replace approximately 1,335 kg of CO_2 emissions, acting as a pilot project towards achieving net-zero carbon emissions.

While higher-income countries have already established over 100 hydrogen refueling systems, simplifying HRS implementation, Nepal faces challenges in areas such as policy development, infrastructure, human resources, and emerging technologies. Although the initial cost will be high, Nepal's abundant hydropower resources will result in lower operational and feed costs for the system. The government should provide subsidies and tax incentives or reductions to facilitate the development of the system and promote carbon-free transportation. Once the initial challenges are overcome, the development of additional systems will become significantly more efficient.

6 ACKNOWLEDGEMENT

We would like to express our deepest appreciation to Mr. Pralhad Gupta and Mr. Nitesh Yadav for their guidance.

REFERENCES

[1] "Global Greenhouse Gas Emissions Data | US EPA." Accessed: Sep. 25, 2023. [Online]. Available: https://www.epa.gov/ghgemissions/global-greenhouse-gas-emissions-data

[2] "Import And Sales." Accessed: Jul. 03, 2024. [Online]. Available: https://noc.org.np/import?type=import&offset=0&max=10

[3] "Nepal's Long-term Strategy for Net-zero Emissions Government of Nepal Kathmandu," 2021.

[4] "Nepal | Climate Promise." Accessed: Jul. 03, 2024. [Online]. Available: https://climatepromise.undp.org/what-we-do/where-we-work/nepal

[5] R. Can Samsun and M. Rex, *Deployment of Fuel Cell Vehicles in Road Transport and the Expansion of the Hydrogen Refueling Station Network: 2023 Update*.

[6] "Statistics - H2Stations.org." Accessed: Jul. 05, 2024. [Online]. Available: https://www.h2stations.org/statistics/

[7] "Green Hydrogen Refueling in Hydrogen Car Started in Nepal - Nep Electrics." Accessed: Jul. 03, 2024. [Online]. Available: https://nepelectrics.com/green-hydrogen-refueling-started-in-nepal/

[8] J. Kurtz, S. Sprik, M. Peters, and T. H. Bradley, "Retail Hydrogen Station Reliability Status and Advances," *Reliab Eng Syst Saf*, p. 106823, Jan. 2020, doi: 10.1016/J.RESS.2020.106823.

[9] "William Robert Grove: Pioneer In Hydrogen Study And Fuel Cells - FuelCellsWorks." Accessed: Jul. 04, 2024. [Online]. Available: https://fuelcellsworks.com/news/william-robert-grove-pioneer-in-hydrogen-study-and-fuel-cells/

[10] "The History of Hydrogen Fuel - KDC Projects." Accessed: Jul. 04, 2024. [Online]. Available: https://www.kdcprojects.com/the-history-of-hydrogen-fuel/

[11] "World's First Hydrogen Fuel Station for Trucks Marks Milestone - Green Hydrogen News." Accessed: Jul. 04, 2024. [Online]. Available: https://energynews.biz/worlds-first-hydrogen-fuel-station-for-trucks-marks-milestone/

[12] "The First Element - Revolutionising green hydrogen." Accessed: Jul. 04, 2024. [Online]. Available: https://thefirstelement.com/

[13] "Press Release." [Online]. Available: https://www.h2stations.org/press-releases/

[14] B. S. Thapa, B. Neupane, H. seong Yang, and Y. H. Lee, "Green hydrogen potentials from surplus hydro energy in Nepal," *Int J Hydrogen Energy*, vol. 46, no. 43, pp. 22256–22267, Jun. 2021, doi: 10.1016/j.ijhydene.2021.04.096.

[15] A. Zhou, W. Zhou, and P. Manandhar, "A Study On The Prospect Of Hydropower To Hydrogen In Nepal," 2020.

[16] R. Bhandari and S. Subedi, "Evaluation of surplus hydroelectricity potential in Nepal until 2040 and its use for hydrogen production via electrolysis," *Renew Energy*, vol. 212, pp. 403–414, Aug. 2023, doi: 10.1016/J.RENENE.2023.05.062.

[17] "Switch Mobility e1 Bus Price, Specifications, Mileage & Images | TrucksBuses.com." Accessed: Sep. 25, 2023. [Online]. Available: https://www.trucksbuses.com/buses/electric/switch-e1-electric-bus

[18] Ł. Wargula, B. Wieczorek, and M. Kukla, "The determination of the rolling resistance coefficient of objects equipped with the wheels and suspension system – results of preliminary tests," *MATEC Web of Conferences*, vol. 254, p. 01005, 2019, doi: 10.1051/matecconf/201925401005.

[19] "Solve friction problems," 2011. [Online]. Available: www.fsmq.org

[20] Marquez Ed and Hahn Christoph, *Vehicle Modeling Using Simulink.*

[21] M. Salazar and M. Vega, "Air Density And Its Air Density And Its Uncertainty Uncertainty."

[22] J. Ruan, P. Walker, and N. Zhang, "Comparison of Power Consumption Efficiency of CVT and Multi-Speed Transmissions for Electric Vehicle," 2018.

[23] "Determining Electric Motor Load And Efficiency Load Ranges: Acceptable Short-Period Acceptable Operating Optimum."

[24] "Tata LP 1112 bus specifications | TrucksBuses.com." Accessed: Jan. 15, 2024. [Online]. Available: https://www.trucksbuses.com/buses/chassis/school-staff/tata-lp-1112/specifications

[25] A. Sadat and M. N. Sadat, "Fuel Cell Efficiency, Voltage Analysis And Performance Optimization," 2014. [Online]. Available: http://www.novapdf.com

[26] "x-engineer.org – Engineering Tutorials." Accessed: Jan. 15, 2024. [Online]. Available: https://x-engineer.org/

[27] "Tata LPO 1625 Starbus Fuel Cell Price - LPO 1625 Starbus Fuel Cell Mileage, Specs & Load Capacity." Accessed: Jul. 03, 2024. [Online]. Available: https://buses.cardekho.com/buses/tata/lpo-1625-starbus-fuel-cell

[28] "Water treatment for green hydrogen what you need to know Reprinted from Hydrogen Tech World magazine."

[29] "Hydrogen production Water Treatment - Pure Aqua, Inc." Accessed: Jul. 02, 2024. [Online]. Available: https://pureaqua.com/expertise/water-treatment-applications/hydrogen-production/

[30] "Medium Capacity 9 x 42 1.0 Cu/ft DI Tank | 505-DI0942." Accessed: Jan. 15, 2024. [Online]. Available: https://www.uswatersystems.com

[31] "High Quality Industry Or Energy Of Pem Elektrolizer Water Electrolysis Pem Electrolyzer Cell Oxygen Hydrogen Production Machine - Buy Pem Elektrolizer,Oxygen Generator Machine,Liquid Oxygen Generator Product on Alibaba.com." Accessed: Jan. 15, 2024. [Online]. Available: https://www.alibaba.com

[32] S. Shiva Kumar and V. Himabindu, "Hydrogen production by PEM water electrolysis – A review," Dec. 01, 2019, *KeAi Communications Co.* doi: 10.1016/j.mset.2019.03.002.

[33] "Hydrogen Production: Electrolysis | Department of Energy." Accessed: Jul. 02, 2024. [Online]. Available: https://www.energy.gov/eere/fuelcells/hydrogen-production-electrolysis

[34] "Diaphragm Compressor - Minnuo Compressor." Accessed: Jan. 16, 2024. [Online]. Available: https://mncompressor.com/gas_compressor/diaphragm-compressor/

[35] "Hydrogen too expensive for Trucking? 85% of the cost is getting it to the pump - Energy Post." Accessed: Jul. 02, 2024. [Online]. Available: https://energypost.eu/hydrogen-too-expensive-for-trucking-85-of-the-cost-is-getting-it-to-the-pump/

[36] K. Pratheesh, "Module 5 contd…compressors Multi-stage air compressor."

[37] "3 kW Screw Compressor - Atlas Copco Australia." Accessed: Jul. 02, 2024. [Online]. Available: https://www.atlascopco.com/en-au/compressors/air-compressor/2-18kw-screw-compressor/3kw-screw-compressor

[38] "Hydrogen calculators - Stargate Hydrogen." Accessed: Jul. 03, 2024. [Online]. Available: https://stargatehydrogen.com/hydrogen-calculators/

[39] S. S. Makridis, "Hydrogen storage and compression."

[40] "Hydrogen storage – Hystorsys AS." Accessed: Jan. 16, 2024. [Online]. Available: https://www.hystorsys.no/mh-storage/

[41] "Hydrogen Tanks." Accessed: Jul. 03, 2024. [Online]. Available: https://hyfindr.com/en/shop/products/category/components/hydrogen-tanks

[42] "Hydrogen Dispenser | Nel Hydrogen." Accessed: Jan. 16, 2024. [Online]. Available: https://nelhydrogen.com/product/hydrogen-dispenser/

[43] H. Barthélémy and Allidières, "Gaseous Hydrogen Refueling Stations: Selection Of Materials For High Pressure Hydrogen Fueling Connectors."

Fault Detection in Turbines Using Machine Learning: A study of the capabilities of Various Classification Algorithms

Aditi Baral[1], Neha Verma[1], Image Adhikari[1], Sailesh Chitrakar[2], Ole Gunnar Dahlhaug[3]

[1] Department of Computer Science and Engineering, Kathmandu University, Nepal
[2] Department of Mechanical Engineering, Kathmandu University, Nepal
[3] Department of Energy and Process Engineering, Norwegian University of Science and Technology, Norway

Email: sailesh@ku.edu.np and aditibaral0819@gmail.com

Abstract. This paper presents a comprehensive study of application of different machine learning techniques for the prediction of turbine faults in an early stage. As the need to ensure an optimal turbine operation through early stage fault detection grows, precise predictive models using machine learning techniques are also increasing rapidly. This study primarily focuses on development and evaluation of the predictive models capable of anticipating fault phases using different range of classification algorithms.It assesses the effectiveness of different classification methods including Support Vector Machine (SVM), K-Nearest Neighbors (KNN), Random Forest, and Convolutional Neural Networks (CNNs) to predict distinct turbine phases.These algorithms are evaluated on experimented dataset encompassing various stages of faults. The results from the algorithm show that CNN was accurately predicting the particular turbine fault phases achieving 80% accuracy with a small length of time series. The deep learning capabilities of CNN capture intricate spatial patterns in turbine data enabling them to discern fault indicators eluded by traditional machine learning methods. The insights gained from this research helps to extend the application of machine learning approaches in the field of turbine fault detection, paving the way for more precise and dependable early failure detection systems.

Keywords: K-Nearest Neighbors(KNN), Random Forest, Convolutional Neural Networks(CNN), Fault detection, Turbines, Support Vector Machine (SVM)

1. Introduction

Turbines are indispensable components in many industrial applications, playing an important role in electricity generation, especially in hydroelectric power plants. The safe and reliable operation of these turbines is very important, because unexpected failures can lead to significant downtime, increased maintenance costs, and even safety risks. The smooth functioning of turbines relies on traditional methods such as scheduled maintenance. Although these measures have provided evidence of reliability, they are often laborious, time-consuming, and fail to provide the accuracy necessary for early fault detection.

Recent years have seen a transformative shift in turbine fault detection, with the integration of machine learning techniques such as Support Vector Machine (SVM), K-Nearest Neighbor (KNN), Random Forest (RF), Convolutional Neural Network (CNN) emerging as a promising solution to overcome the limitations of traditional methods [1]. Machine learning techniques in fault detection are helpful to predict the failure of the turbine by taking as inputs the various sensor data and feeding them into a model. This highlights the shift towards data-driven methods emphasis, ensures optimal turbine conditions by detecting faults in a timely manner, thereby reducing downtime and improving efficiency.

This paper begins with a review of the application of machine learning techniques,including SVM, KNN, CNN and RF, aimed at identifying turbine faults at their onset. While there are others who have worked on similar problems in the past, most of their approaches have considered data from a single time instance which is not able to accurately predict the condition of the turbine. For instance, Garcia et al. (2006) applied MLP to wind turbine gearbox monitoring, using indicators like bearing and oil temperature, generate power, and fan speed [2]. Schlechtingen and Santos (2011) compared ANN and autoregressive models for wind turbine fault identification, utilizing temperatures of critical components and shaft speed as inputs [3]. Additionally, Saeed et al. (2013) employed PCA for feature extraction to predict crack length on hydro turbine blades [4].

This paper focuses on investigating the accuracy of machine learning models to predict the fault in turbines in the early stage. By considering data from multiple time instances and using classification models on these series of data, the accuracy of these models in predicting faults in turbines is evaluated. The emphasis on the temporal evolution of turbine data enables a more comprehensive and effective solution to early fault prediction in this work.

2. Materials And Methods

This section outlines the core materials and methodologies employed in our study, which is centered on the early detection of turbine faults using machine learning algorithms. This research focuses on comparing the effectiveness of various algorithms and techniques aimed at predicting the faults in turbines. One effective way of doing this is by analyzing the data from different pressure and vibration sensors and predicting the phase of the turbine from previous labeled data of sensor readings and their respective phases.

The sensor readings from pressure sensors (PGV1, PGV2, PGV3) and vibration sensors(ATB1, ATB2) have been used for this research. These sensor data were fed into algorithms including Support Vector Machines(SVM), K-Nearest Neighbors (KNN), Random Forest (RF), and neural networks like Convolutional Neural Network (CNN), and Fully Convolutional Network (FCN). The data in this research consisted of that collected under five operating points - Deep Part Load (DPL), Part Load (PL), Best Efficiency Point (BEP), High Load (HL), and Full Load (FL).

2.1 Data Collection and Preprocessing

The data for this study was provided by Norwegian University of Science and Technology (NTNU) [5], which was collected from a Francis turbine located within the laboratory of the NTNU. These turbines were subjected to controlled experimentation to induce crack growth in the turbine blades, simulating realistic damage scenarios. These experiments involved initiating cracks at the trailing edge of a blade, which followed a semi-elliptical path, eventually leading to the detachment of a shark-bite-shaped fragment [5]. The crack was systematically extended through eight stages, with specific lengths detailed in Table 1, and an additional stage representing an enlarged opening as shown in Figure 1.

Stage 0 signifies the initial or normal condition of the turbine blades, serving as a baseline for comparative analysis throughout the experimentation stages. At this stage, the blades are free from induced damage. On the other hand, Stage 9 represents the culmination of the experiment, wherein the turbine blade is entirely detached, resulting in a distinct shark-bite-shaped fragment.

Table 1 shows the lengths of cracks that were extended in 8 stages from the normal situation. Figure 1 shows the detailed view of the crack propagation stages. In the figure, stage 0 represents the normal condition of the turbine. Stages 1 through 9 show the damaged blade. It is to be noted that stage 4-7 represents an unnatural crack development but they were still included to investigate whether a potential fracture would extend and detach a fragment during the measurements.

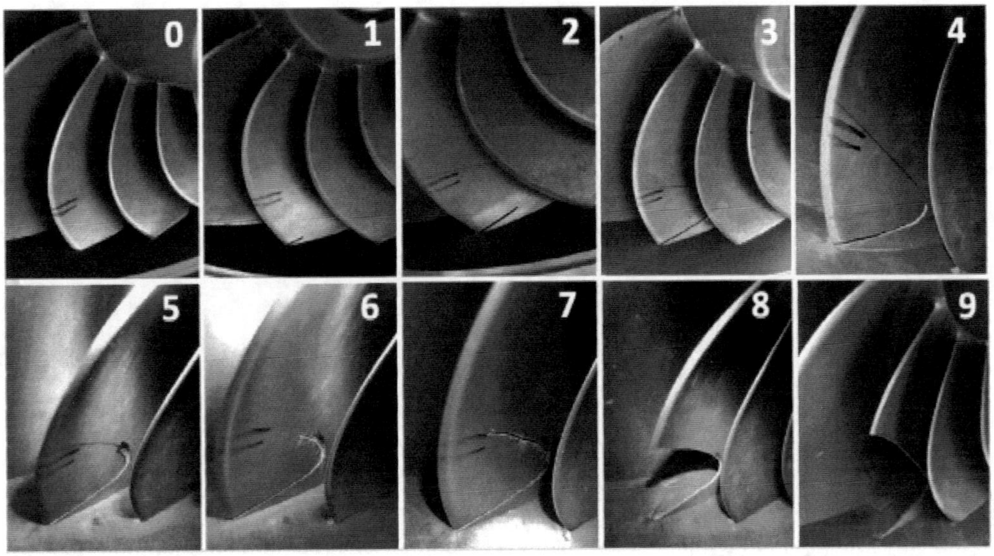

Figure 1. Crack propagation stages: 0 indicates normal condition, 1-7 show the crack growth, 8 indicate a fully realized detached fragment and 9 shows an enlarged opening of the detached fragment [5]

Table 1. Crack Propagation Stages and its total crack length [5]

Stage	Crack Length [mm]
1	15
2	30
3	50
4	72
5	95
6	115
7	140
8	170

The data for this study was acquired through sensors placed on different parts of the turbines. Figure 2 provides a spatial representation of the pressure sensors and accelerometers used during the experiments. Pressure sensors were strategically placed in the stationary domain of the turbine, while accelerometers were positioned at key locations to capture vibration data essential for the analysis of turbine behavior.

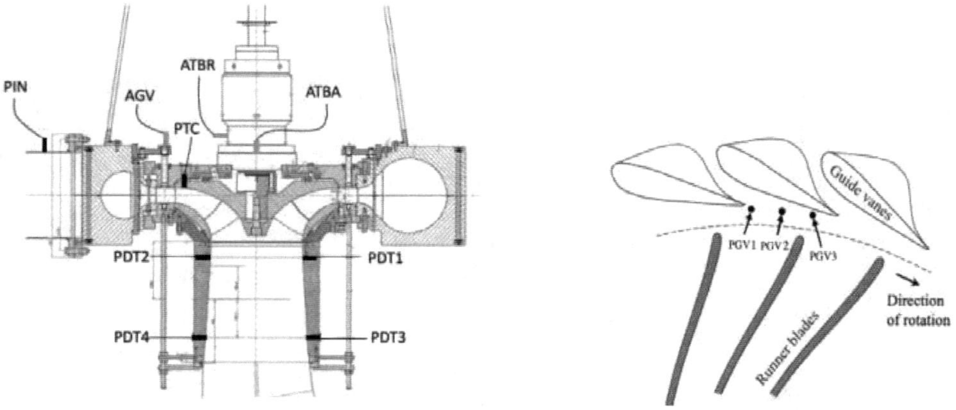

Figure 2. Location of the pressure sensors and accelerometers used for the experiment [5]

The datasets contained a variety of sensor data recorded at different heads as shown in Figure 2. However this research specifically focussed on analyzing the data from the pressure sensors (PGV1, PGV2, PGV3) and vibration sensors (ATB1, ATB2) for the fault detection models.

Initially, the raw data was available as binary files (.mat and .tdms file formats). The .tdms files contained sensor properties, while the .mat files held the sensor readings and data. To facilitate easier interpretation and visualization, these files were converted into .csv format. This conversion allowed for more accessible and efficient handling of the data, enabling better integration with various data analysis tools and frameworks.

The data was then organized into different folders corresponding to each phase of the turbine, based on the sensor data. Each folder contained .csv files representing different head or operating points of the turbine. This structured organization resulted in tabular data that included sensor names and their readings at various operating points, making the data more manageable and analyzable.

This preprocessing step ensured that the data was clean, well-structured, and ready for analysis. By converting the data into a consistent format and organizing it systematically, it was possible to streamline the subsequent data analysis process, thereby enhancing the accuracy and reliability of the predictive models. The preprocessing choices made in this study are crucial for validating the findings and ensuring the robustness of the predictive maintenance system.

2.2 Machine Learning Algorithms

This research focussed on analyzing and exploring different machine learning algorithms that can provide fine accuracy to turbine fault detections. The objective is to provide a comprehensive understanding of different classification algorithms and their performance in the context of early fault detection.

2.2.1. Support Vector Machine (SVM)

Support Vector Machines (SVMs) are a valuable class of algorithms known for their ability to create distinct boundaries between different data classes, often achieved through the transformation of data into higher dimensions and the identification of an optimal hyperplane [6]. In this research, SVMs were applied to a dataset derived from five sensors, including three pressure and two vibration sensors.

To facilitate SVM analysis, the data was initially preprocessed by merging two operational phases each into a single category. Given the substantial volume of data recorded by different sensors across 10 operational phases and minimal differences observed between many of the phases, the classification task became more manageable and effective for the SVM. For instance, Phases 1 and 2 were grouped into class 1, phases 3 and 4 into class 2, representing normal operating conditions and phases 5 and 6 into class 3 and phase 7 and 8 into class 4, indicating potential failure conditions. Additionally phases

9 and 10, representing more failure conditions, resulted in five classes in total. This strategic grouping allowed the SVM to create clearer boundaries and enhance anomaly detection.

Following preprocessing, data normalization was performed to ensure that all features contributed equally to the model's decision-making process. The SVM model was then trained on data from groups representing normal and early-stage fault conditions, such as Group 1 and Group 2, and subsequently tested on groups representing later stages of fault development, such as Group 3, Group 4 and Group 5. This approach exposed the SVM to a broad range of conditions, thereby classifying data points into one of the five classes. This approach showcased the potential of SVMs for early-stage fault detection in turbine systems, while acknowledging the need for tailored methodologies in different scenarios and datasets to achieve desired accuracy and effectiveness.

2.2.2. K- Nearest Neighbors (KNN)
K-Nearest Neighbor (KNN) is the simplest of the data driven algorithms where it determines the label of the test data by taking a vote among its k nearest neighbor and picking the label with highest vote. It stores the training data during the training phase which leads it to have high storage requirements and also makes it more computationally intensive during the prediction. This also results in KNN not scaling well with large datasets.

Dynamic time warping was used as the distance metric to determine the nearest neighbor [7]. Instead of measuring the euclidean distance between the corresponding points in the time series, DTW tries to align the time series and give us a cost of aligning. The lower the cost, the more similar the signals. This was run on pressure sensor (PGV1) data of BEP. The data was divided into multiple series of 915 points as the lowest points in phase 10 of PGV1 were roughly 915 points away. 2000 of such series were used for training, 200 from each phase. It was tested for 500 series of the same length for k=20, 25 and 30.

2.2.3. Decision Trees
Decision trees is a machine learning algorithm which makes use of tree structure where each internal node denotes a test on an attribute, each branch represents an outcome of the test, and each leaf node (terminal node) holds a class label. Random forest is an ensemble machine algorithm which builds decision trees on different samples and takes their majority vote to predict the label [8].

Time Series Forest Classifier adapts the random forest classifier to series data which was used to classify the sensor data. It splits the series into random intervals (sqrt(series_length) intervals), with random start positions and random lengths and extracts summary features (mean, standard deviation, and slope) from each interval into a single feature vector [9]. The decision trees are trained on these extracted features and are added to the ensemble.

The number of decision trees used in the forest i.e n_estimators were taken to be 50, 100 and 500 and were run on pressure sensor (PGV1) data of BEP (Best Efficiency Point). The length and number of time series taken were similar to the KNN.

2.2.4. Convolutional Neural Networks (CNNs) and Fully Convolutional Network (FCN)
Convolutional Neural Networks (CNNs) are a class of neural networks mostly used in image processing. They make use of convolution operations between the input and the filter which are learned from training examples to extract the features [10]. These mostly contain 3 layers: Convolution Layer, Pooling Layer and Fully Connected Layer.

A Fully Convolutional Network (FCN) in a neural network consists of convolutional operations applied to the input data. Unlike traditional fully connected layers, where each neuron connects to all neurons in the previous layer, convolution operations use learnable filters across the input data, preserving spatial information, as shown in Figure 3.

In the convolution layer, the filter slides through the input and takes the dot product. Pooling layer is used for downsampling and to reduce the size of input (and thus the computation). In Fully Connected Layer every input is connected to every output by a learnable weight.

For this research, the first convolutional neural network was implemented with the same amount of data as other algorithms which were first normalized. Three convolution layers were used with kernel size 8, 5, 3 and 128, 256, 128 filters respectively followed by a fully connected layer. These configurations allowed the models to effectively capture patterns of varying granularity in the sensor data, enhancing feature extraction. Spatial dropout and batch normalization were added after each convolution layer. Spatial dropout was applied to mitigate overfitting by randomly dropping units during training, thus promoting generalization. Batch normalization was incorporated after each convolution layer to stabilize and accelerate training by normalizing the activations, which also helps in reducing the internal covariate shift.

The data from pressure sensors (PGV1, PGV2, PGV3) and vibration sensors (ATB1, ATB2) were then used to train the model for all the operating points and the results were observed. The length of the time series was decreased to a significantly lower number by considering only 20 points of each sensor to see how the model performs for smaller time series.

The training parameters for optimization were configured with the Adam optimizer, which included a learning rate of 0.001. For optimization, the Adam optimizer was selected due to its adaptive learning rate capabilities, enhancing convergence speed and accuracy. A learning rate of 0.001 was set to balance between convergence speed and the risk of overshooting the minimum loss. Categorical Cross-Entropy was used as the loss function, suitable for multi-class classification tasks, and accuracy was chosen as the evaluation metric to assess the model's predictive performance. The Fully Connected Layer in CNNs was replaced by Global Average Pooling in FCNs to reduce the number of parameters and maintain spatial hierarchies, aiding in preventing overfitting and ensuring a robust model[11]. These hyperparameters collectively contributed to the models' ability to discern intricate patterns and anomalies in the turbine data, making them effective for early fault detection.

Figure 3 shows a schematic representation of a Fully Convolutional Network (FCN). The diagram outlines the sequential layers involved in the processing of input data to produce an output. It begins with an Input Layer, followed by Convolution Layer 1, Batch Normalization Layer, Convolution Layer 2, another Batch Normalization Layer, and then a second Convolution Layer. The data is then processed through a Global Average Pooling Layer before reaching the final Output Layer.

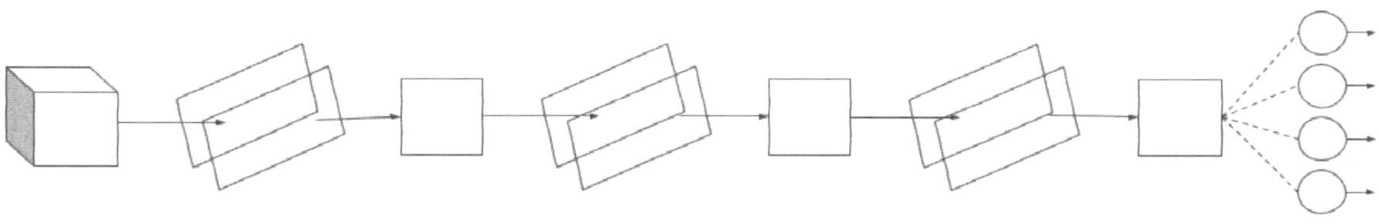

Figure 3. Fully Convolutional Network

3. Results and Discussion

This research involved the application of various machine learning algorithms including SVM, KNN, Random Forest, CNN and FCN to predict faults in turbines. Each of the aforementioned algorithms exhibited distinct performance characteristics, having their own set of advantages and shortcomings.

3.1 SVM Model Performance

It is noteworthy that the SVM model did not incorporate temporal dependencies within the dataset, i.e. data from multiple timestamps were not considered, which has had a discernible impact on the model's overall accuracy in comparison to other implementations. Specifically, the SVM model yielded an accuracy score of 0.1451 falling within the range of 14% to 20%, indicating a significant disparity when benchmarked against alternative methodologies.

3.2 KNN Model Performance:

The K-Nearest Neighbors (KNN) algorithm demonstrated variable performance based on the choice of the hyperparameter k. KNN achieved an accuracy of 50.2% when k was set to 20 and 25 and an accuracy of 49.6% for k=30. Furthermore, KNN was particularly slow during the testing as finding the closest neighbor was done during testing. Furthermore, dynamic time warping also requires high computation which worsens this problem. Optimization methods like LB_Keogh [12] and prunedDTW [13] can help to improve the performance. Table 2 shows the phase predicted by KNN for those three values of k and the actual phases.

Table 2. Result of KNN for k=30, 25 and 20

Actual Phase	**Predicted Phase**		
	P_30	**P_25**	**P_20**
Phase 1	Phase 2	Phase 2	Phase 2
Phase 1	Phase 1	Phase 1	Phase 1
Phase 1	Phase 2	Phase 2	Phase 2
Phase 1	Phase 1	Phase 1	Phase 1
Phase 1	Phase 1	Phase 1	Phase 1
...	...	...	...
Phase 9	Phase 7	Phase 7	Phase 9
Phase 9	Phase 5	Phase 8	Phase 8
Phase 9	Phase 9	Phase 9	Phase 9
Phase 9	Phase 9	Phase 9	Phase 9
Phase 9	Phase 5	Phase 5	Phase 7

3.3 Time Series Forest Performance:

Using time series forest, it was observed that the accuracy increased up to 66%. Moreover, the accuracy remained relatively stable with an increase in the number of n_estimators. It had a high number of wrong predictions for the earlier two stages but had good results for the final two stages. Correct prediction per phase also remained very similar with the increase in n_estimators. Extending time series forest to handle data from multiple sensors can be considered to improve the performance. Other tree based algorithms like Random Interval Spectral Ensemble (RISE), which unlike time series forest uses Spectral features, can be implemented to get better accuracy [14].

IOP Conf. Series: Materials Science and Engineering **1314** (2024) 012004 doi:10.1088/1757-899X/1314/1/012004

3.4 CNN and FCN Performance:

While using CNN with spatial dropout and batch normalization for the same data, training accuracy observed after training for 20 epochs was high. This however decreased to a lower amount during testing which was the condition of overfitting. With FCN the training accuracy decreased but the test accuracy was observed to be much closer to the training accuracy than with CNN.

The models were trained using various sensor data at various operating points. FCN consistently demonstrated closer alignment between training and testing accuracy across these variations.

The table shows the accuracy achieved by CNN and FCN during training and testing time and the sensors and operating point of which the data is taken. The training for FCN was done for 10 epochs in cases where it is not otherwise stated and for CNN it was for 20 epochs.

Table 3. Results from CNN and FCN with various amount of data in percentage (rounded down to integer value)

Sensors	**Operating Points**	**Convolutional Neural Network (CNN)**		**Fully Convolutional Network (FCN)**	
		training	**testing**	**training**	**testing**
PGV1	BEP	82	52	54*	50*
PGV1, PGV2	BEP	88	56	68*	40*
PGV1, PGV2	BEP	88	56	82	73
PGV1, PGV2, PGV3	BEP	89	63	85	80
PGV1	BEP, DPL	52	32	40 (50 epoch)	36
PGV1, PGV2, PGV3, ATB1, ATB2	BEP, DPL	85	79	97	97
PGV1, PGV2, PGV3, ATB1, ATB2	BEP, DPL, PL, HL, FL	91	87	98	98
PGV1, PGV2, PGV3, ATB1, ATB2	BEP, DPL, PL, HL, FL			85 (length 20)	83

*FCN with spatial dropout layer

It was observed that FCN performed better without the dropout layer, so it was removed after including 2 sensors.

3.5 Confusion Matrix Insights

The confusion matrix is used to visualize the result by showing the actual phase and predicted phase for FCN with a time series with length 20. In the confusion matrix it was observed that the majority of the series were classified correctly as indicated by the brightly colored diagonal. Aside from that, the highest misclassification is the series from phase 9 being classified as phase 10. The reverse of this also has the number on the higher end.

Some of the other common misclassifications were series of phase 1 and 2 being classified as 5, phase 5 being classified as 1, phase 7 being classified as phase 2 and 5. Phase 7 has the lowest number of correct classification, all of which can be seen in the confusion matrix in Figure 4.

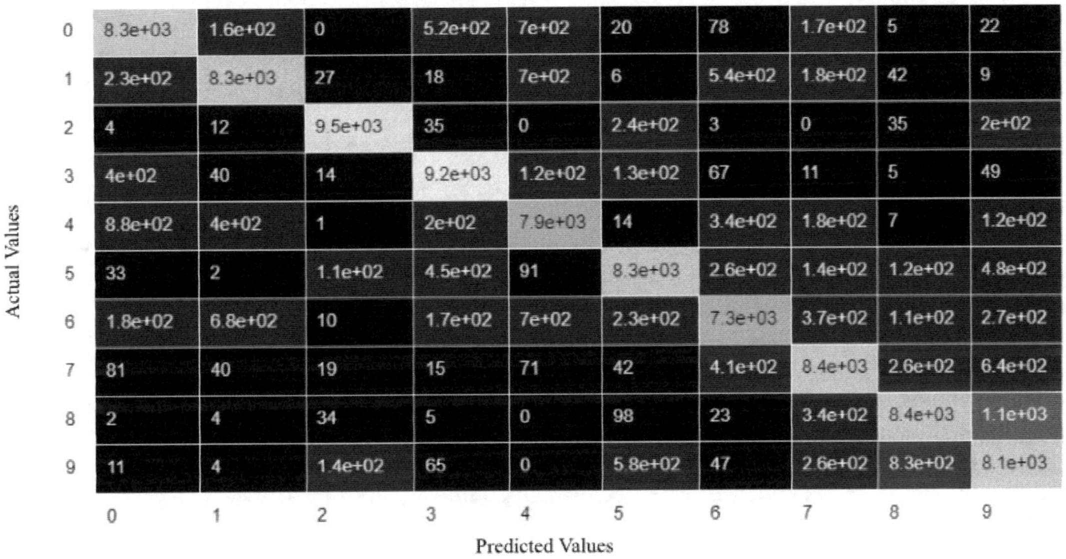

Figure 4. Confusion matrix for FCN using length 20

Note that the number in the matrix is one less than the actual corresponding phase as it is marked as 0-9 instead of 1-10.

In summary, this study highlighted the varying performance of machine learning algorithms in turbine fault prediction. SVM showed limitations when not considering temporal dependencies, while KNN, Random Forest, and CNN exhibited promising results with different strengths and weaknesses. So, the choice of algorithm should be carefully considered in the context of specific application requirements and dataset characteristics.

4. Conclusions

In conclusion, this research has demonstrated the valuable potential of utilizing sensor readings from turbines to predict maintenance time. Data from pressure and vibration sensors offer substantial insights into the condition of turbine blades, enabling the development of machine learning models for blade health assessment and wear estimation. While classification algorithms such as SVM and KNN exhibit applicability in certain scenarios, deep learning approaches, particularly CNN-based methods, emerge as notably more efficient for predicting blade wear levels and turbine phase classification.

Quantitatively, the findings indicate that the CNN-based methods have been able to achieve high accuracy levels in predicting turbine blade conditions. The accuracy score from previously implemented algorithms to the CNN based approach produced a growth from around 20% to well over 80% which indicates a robust model. However, it is imperative to note that the model's evaluation has primarily been conducted on data acquired in controlled settings, necessitating further validation on real-world datasets. Future research should also focus on fine-tuning model parameters to enhance robustness and adaptability across diverse turbine types. These quantitative insights underscore the potential practicality and effectiveness of our approach in the domain of turbine maintenance prediction.

4.1 Limitations

This study provides valuable insights into fault detection of turbines using various machine learning algorithms, applied to controlled experimental data. However, several limitations must be acknowledged while training the models.The dataset used to train the model were based on sensor data from the experimented setup in this research, lacking the diverse range of operating conditions and fault scenarios which is encountered in real-world turbine operations, potentially which could introduce biases and limit the model's generalizability. Moreover, artificially induced faults may not accurately replicate the complex dynamics and nuanced fault indicators present in operational turbines, thereby affecting the algorithms ability to effectively classify and predict real-world fault patterns. Additionally, variations in experimental setup, such as sensor placements and environmental conditions, pose challenges to the algorithms scalability and robustness in practical applications. To address these limitations, rigorous validation with real-world turbine data is required to ensure the models can reliably detect faults under diverse operational conditions and scenarios.

4.2 Future Enhancement

Future research should be done to focus on exploring and evaluating more deep learning architecture and techniques to enhance the fault detection in the turbine systems. Also, different deep learning techniques should be combined into hybrid models as the single model could not accurately capture all the intricate patterns of the sensors like integrating Convolutional Neural Networks (CNNs) with Long Short-Term Memory (LSTM) networks, could harness the strengths of both architectures, leading to more robust fault detection model. Besides this, different transformer based models could be analyzed for their ability to capture the long term dependencies and complex patterns in sensor data. Meanwhile, the incremental learning approach could be useful for real world scenarios. In this approach, the model is continuously updated as new data arrives, enabling it to adapt to new patterns and maintain high performance in a dynamic environment ensuring that the model remains effective and accurate.

Acknowledgements

The project was funded by the EnergizeNepal program with the project ID: TTL-Fault (ENEP-RENP-II-23-03) registered at Kathmandu University.

5. References

[1] Black I M, Richmond M, & Kolios A (2021, March 11). Condition monitoring systems: a systematic literature review on machine-learning methods improving offshore-wind turbine

operational management. *International Journal of Sustainable Energy*, *40*(10), 923–46. https://doi.org/10.1080/14786451.2021.1890736

[2] Garcia M C,Sanz-Bobi M A, Pico J D, SIMAP: Intelligent System for Predictive Maintenance: Application to the health condition monitoring of a wind turbine gearbox, Comput. Ind. 57 (2006) 552–68.

[3] Schlechtingen M, Santos IF, Comparative analysis of neural network and regression based condition monitoring approaches for wind turbine fault detection, Mech. Syst. Signal Process. 25 (2011) 1849–75.

[4] Saeed R A, Galybin A N, Popov V, 3D fluid–structure modeling and vibration analysis for fault diagnosis of Francis turbine using multiple ANN and multiple ANFIS, Mech. Syst. Signal Process. 34 (2013) 259-76. doi:https://doi.org/10.1016/j.ymssp.2012.08.004.

[5] Støren GK, (2021). *Signature investigation of typical faults on a Francis turbine* [Master thesis, NTNU]. NTNU Open https://ntnuopen.ntnu.no/ntnu-xmlui/handle/11250/2779679

[6] Widodo A & Yang B S. (2007, August). Support vector machine in machine condition monitoring and fault diagnosis. *Mechanical Systems and Signal Processing*, *21*(6), 2560–74. https://doi.org/10.1016/j.ymssp.2006.12.007

[7] Tiozzo A (2022, November 9). *Dynamic Time Warping for Time Series Classification | Eni digiTALKS*.Medium. https://medium.com/eni-digitalks/dynamic-time-warping-for-time-series-classification-6482cf9d7c3b

[8] Deng, Houtao, et al. "A Time Series Forest for Classification and Feature Extraction." *arXiv.Org*, 18 Feb. 2013, arxiv.org/abs/1302.2277.

[9] Surucu O, Gadsden S ., & Yawney J (2023, July 1). *Condition Monitoring using Machine Learning: A Review of Theory, Applications, and Recent Advances*. Expert Systems With Applications; Elsevier BV. https://doi.org/10.1016/j.eswa.2023.119738

[10] Hamid M A, Ibrahim R A, Abdelgalil M and Desouki H, "Bearing Fault Identification for High-Speed Wind Turbines using CNN," *2023 11th International Conference on Smart Grid (icSmartGrid)*, Paris, France, 2023, pp. 1-5, https://doi: 10.1109/icSmartGrid58556.2023.10171018

[11] Wang Z, Yan, W, & Oates T, (2017). Time Series classification from scratch with Deep Neural Networks: A strong baseline. *2017 International Joint Conference on Neural Networks (IJCNN)*. https://doi.org/10.1109/ijcnn.2017.7966039

[12] Keogh, E., Ratanamahatana, C. Exact indexing of dynamic time warping. Knowl Inf Syst 7, 358–386 (2005). https://doi.org/10.1007/s10115-004-0154-9

[13] Silva, Diego F, and Gustavo. Speeding up All-Pairwise Dynamic Time Warping Matrix Calculation. 30 June 2016, https://doi.org/10.1137/1.9781611974348.94

[14] Lines, Jason, et al. "Time Series Classification with HIVE-COTE." ACM Transactions on Knowledge Discovery from Data, vol. 12, no. 5, 5 July 2018, pp. 1–35, https://doi.org/10.1145/3182382

Experimental Study of Mechanical Properties of Natural Fiber Polymer Composite

V GC [1], R Joshi [1], H Giri [1], S Sujakhu [1]* and M Shah [1]

[1] Department of Mechanical Engineering, Kathmandu University, Dhulikhel, Nepal

E-mail: surendra.sujakhu@ku.edu.np

Abstract. Numerous studies and research projects have been conducted to develop natural fiber polymer composites. The researcher and engineers develop a composite by reinforcing natural fiber with a polymer that exhibits the desired mechanical properties. The experiment determines the composite's mechanical properties, such as tensile, compression, flexural, and impact strength. This paper presents the experimental study of the natural fiber-reinforced composite, where hemp and jute fibers were reinforced with epoxy polymer as the test specimens for the experiment. The primary approach involved fabricating composite samples with fiber-weight fractions and matrix compositions. Then the appropriate sample is experimented with for its tensile strength, compression strength, and impact strength. Another property of the composite's the composite's water absorption was investigated through a water absorbability test. Key results show that composites containing a 30% weight fraction of natural hemp fiber demonstrated promising mechanical properties. This composite showed a tensile strength of 50 MPa, a compression strength of 55.6 N/m^2, an impact strength of 58.9 kJ/m^2, and a water absorbability of 7%. These results may offer valuable insight into the mechanical behavior of natural fiber composites, which can be used to optimize material design and engineering applications. Understanding the limitations and strengths of these composites allows for informed decision-making in selecting appropriate materials for specific engineering projects.

Keywords. Natural fiber, epoxy resin, mechanical property, composite

1. Introduction

Composite material is composed of two or more materials sharing each constituent property. The composite is composed of filler material embedded in the matrix. Normally, fiber is used as filler material, and the matrix could be thermosetting, thermoplastic, and/or elastomer type. The matrix is used to bind the fiber and ensure the orientation of the load [1]. Fibers used in composite preparation are broadly classified as synthetic (glass, carbon, basalt, etc.) or natural, each offering distinct advantages and challenges [2]. These composite offers material property such as strength, lightweight, stiffness, resistance to fatigue, corrosion, and impact strength, making them ideal for the diverse application. They have replaced many metallic components in the various manufacturing sectors. Due to their excellent mechanical properties, these materials have a wide range of applications in various industries such as aerospace and defense, energy, automotive, construction, sport, and many more [3]. However, the synthetic fiber's composite is non-biodegradable and contributes to carbon footprint [4].

In engineering, material selection of a sustainable material plays huge role. One such sustainable solution could be a composite from Natural fiber. The competitive edge natural fiber provides over synthetic fiber as their abundance, availability, and low cost. But their moisture absorption property is key concern for their use in structural applications [5]. In addition to this, the mechanical properties of natural fiber composites are often inferior to those of synthetic fiber composites. Therefore, it is not possible to replace natural fiber with synthetic fiber in all application. However, natural fiber composites present a viable alternative in applications where exceptional strength is not required. Natural fiber composite is now used different application such as floor protection in passenger car, roofing, flooring and insulation, filters in cigarettes, membranes and barriers, biosensors and catalysts, thin sheets and films, packaging materials, etc [6]. Natural fibers obtained from the different sources like plant, animal, minerals, and geological processes can decompose [7]. Normally plant fibers are used for industrial applications.

Fibers such as flax, hemp, bamboo, and jute have been marketed as prospective substitutes for traditional composite reinforcement [8]. Roughly aligning these fibers possesses limitations such as poor matrix interfacial bonding, moisture absorption, and low damage tolerance. To address these issues, fiber is made in textile form (woven or knitted) [9]. The fiber in fabric is arranged in various orientation with respect to the axis of the composite and they are characterized by high in-plane strength and stiffness. The composite is prepared by assembling several fibrous layers and combining them with a matrix [10]. The most widely used matrices for natural fiber composites are thermosetting resins (such as epoxy resin) and thermoplastic resins (such as polypropylene (PP), polyethylene (PE), polyvinyl chloride (PVC), polylactic acid (PLA), etc. Thermosetting resins have higher thermal stability and mechanical strength than thermoplastic resins but are more brittle and difficult to recycle [11].

Several studies have been conducted to understand the mechanical properties of natural fiber composites. Darshil et al. experimented with flax/polyester for the small-scale wind turbine blade and suggested the composite is suitable for use in that case [12]. John et al. examined mechanical properties of bamboo epoxy laminate. The result showed the composite to have tensile strength of 175 MPa and modulus to be 20 GPa[13]. Palanikumar et al. performed mechanical testing for the sisal fiber and glass fiber composite[14]. Sahayaraj et al. investigated on different mechanical property for jute and hemp in proportion 50:50 reinforced with the matrix PLA, and the composite exhibits a tensile strength of 69 MPa, an impact strength of 6.37 J [15].

Similarly, various studies have been done on the different types of natural fibers and matrices. However, there are very few papers in natural fiber composite in the context of Nepal, and resin locally available in Nepal. In this paper, we focus on two types of natural fibers: hemp fiber and jute fiber. We use epoxy resin as the matrix material for all the composites. We fabricate the composites using the hand lay-up method with different fiber weight fractions ranging from 10% to 50%. We test the composites for water absorption taking reference of the ASTM D570 standard [16]; tensile strength with reference to ASTM D3039 standard [17]; compressive strength with reference to ASTM D695 standard [18]; and impact strength with reference to ASTM D256 standard [16]. The fabrication of the composite involved the wet hand layup method. We further analyze the result obtained various mechanical testing and discusses the failure mode as well.

2. Material and Method

The experimental study was conducted to determine the mechanical properties of natural fiber composites by fabricating the composites using the hand lay-up method for different matrix weight fractions ranging from 10% to 50%. The composites were then tested for water absorption, tensile strength, compressive strength, and impact strength. The effects of fiber type, fiber content and water absorption on the mechanical properties of the composites were analyzed. Possible failure mechanisms and factors influencing the mechanical behavior of the composites are suggested. The following sub-sections give further details about material selections and experimentation.

2.1. Material

2.1.1.Fiber

In this study, jute and hemp fibers were taken for experimentation purposes (fabric properties mentioned in Table 1). Additionally, a flax and cotton fabric was used as a reference to compare the mechanical behavior of the jute and hemp composites. The fabric taken is a woven mat having a biaxial orientation. The fabric is cut into 10*10 cm to observe the following properties of the fabric.

Table 1. Properties of the fiber

Fibers	Weight (g)	Mesh size (mm)	No of fibers in warp direction	No of fiber in weft direction
Hemp	4.60	0.28	66	54
Jute	3.45	0.76	43	39

2.1.2.Matrix

A thermosetting epoxy resin (318A-7T) of the brand Sparko was used as a matrix material for the composite (properties mentioned in Table 2). The resin was mixed with hardener (318B-7T) in the ratio of 3:1 by volume and 100:33 by weight. They are bought from the authorized distributor of the Sparko adhesive Champak and Chirag International. The curing time of the epoxy resin was 24 hours at 25° C. The following are the properties of the resin by manufacturer:

Table 2. Properties of the resin before hardening

Part	318A-7T	318B-7T
Color	Transparent	Transparent
Specific gravity	1.15	0.97
Viscosity (25° C)	2000-4000CPS	50max CPS
Mixing ratio	A:B = 100:33 (weight ratio)	
Hardening condition	At 25° C 24-48 hours (100g)	
Usable time	120 min 25° C (100g)	

2.2. Sample Preparation

Different details of the experimental setup are given in the sub-sections below:

2.2.1.Mold Preparation

Mold preparation involved the use of wood, foam, plastic paper, and candle wax, ensuring the uniformity of the composite specimens. Various tools, including weighing machines, scissors, brushes, and more, were employed during the fabrication process.

2.2.2.Specimen Composition and Dimension

Composite specimens were prepared with varying fiber-to-resin weight ratios (20/80, 30/70, 40/60, 50/50, 60/40) to determine the optimal proportion where the fiber and matrix properly bind. After the observation, the appropriate proportion is used for the preparation of specimens for different tests. The following table shows the labeling of specimens:

Table 3. Label of different composite specimen

Label of specimen	Fiber type	Proportion of matrix by weight	Proportion of fiber by weight
H73	Hemp	70%	30%
H82	Hemp	80%	20%
J73	Jute	70%	30%
J82	Jute	80%	20%

2.2.3. Composite Fabrication

The composite specimens were fabricated using the hand layup technique. This method involves a systematic process of mold preparation, precise cutting and preparation of the fiber, accurate weighing of the fiber and resin, thorough mixing and application of the resin, molding of the composite, and finally, cutting the specimens to standardized dimensions.

Figure 1. Sample Fabrication

2.2.4. Experimental Testing of the Specimens

- *Tensile Test:* Tensile tests were conducted on AIM UTM machine. The specimen is prepared with reference to the ASTM D3039, where the size is about 250*25*3.2 ± 2.5 mm. For gripping sand paper was wrapped at the top of specimen where it is clamped.
- *Compressive Test:* The compressive strength test was prepared according to the ASTM D 695 having dimension 25.4*12.7*12.7 mm. The test was done in a compressive strength testing machine named ENKAY DIGIMAX- 109.
- *Charpy Impact Test:* The standard size of the sample for the Charpy impact test is 55x10x10 mm and has V-shaped notch. The specimens were prepared according to standard; however, maintaining a consistent thickness was challenging. Therefore, specific impact energy was calculated to provide an accurate representation of the material properties.
- *Water Absorbability Test:* The water absorbability tests involved submerging specimens in water-filled containers for 36 hours with meticulous observation and quantification of water absorption. The size of test specimen was 20*15*4 mm.
- *Microscopic Study:* A Radical stereo zoom microscope was used to facilitate a high-resolution examination of composite specimen. External camera is used to capture the microscopic view. The microscopic view of composite is used to analyze the void formed in composite and failure mode after experimentation.

IOP Conf. Series: Materials Science and Engineering **1314** (2024) 012005 doi:10.1088/1757-899X/1314/1/012005

3. Result and Discussion

3.1. Initial Observation of the Composite

The composite of the hemp and jute was prepared first with the following resin-to-fiber weight ratios 40/60, 50/50, 60/40, 70/30, and 80/20. Preliminary observation showed that 40/60 and 50/50 ratios did not achieve desired adhesion. Consequently, the composites with weight proportions of 70/30 and 80/20 were selected for the mechanical testing. The figure below depicts the specimen having improper adhesion:

Figure 2. Composite specimen with improper adhesion

3.2. Tensile Test

The average tensile strength of the specimens is different for different compositions of matrix and fiber. H73 exhibits the highest tensile strength at 50.00 MPa, and J82 showed the lowest at 32.96 MPa. The specimens H82 and J73 had tensile strengths of 41.50 MPa and 36.63 MPa, respectively. The results give an insight that increasing the proportion of epoxy resin lower the tensile strength. The standard deviation values provide insight into the consistency and reliability of the tensile strength measurements. The greater variation in H82 implies that there must have been some error while preparing the composite specimen. The result of the tensile test experiment of composite materials with different fibers and compositions are illustrated in Figure 3 and 4 (presented in Table 4).

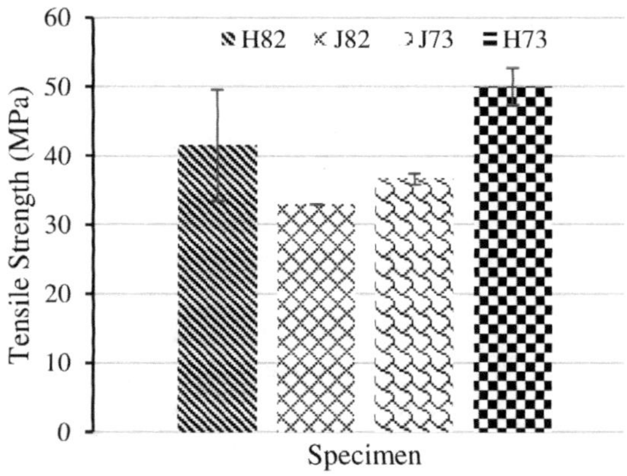

Figure 3. Tensile strength of composites

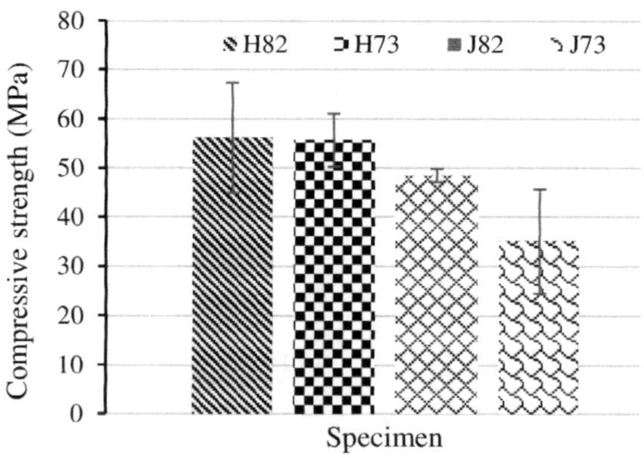

Figure 4. Compressive strength of composites

Table 4. Tensile strength, compressive strength and impact energy result of the composites

Specimen	Tensile Strength (MPa)		Compressive Strength (MPa)		Impact Energy (kJ/mm^2)	
	Average	SD	Average	SD	Average	SD
H82	41.50	8.04	56.15	11.19	32.74	8.62
H73	50	2.64	55.65	5.36	58.87	20.46
J82	32.96	0.03	48.56	1.35	34.05	1.43
J73	36.63	0.8	35.12	10.65	50.28	10.01

3.3. Compression Test

It is observed that compressive strength for H82 and H73 composite specimens have the highest compressive strengths, with values of 56.15 N/mm² and 55.65 N/mm², respectively. The J82 specimen follows with a compressive strength of 48.56 N/mm², while the J73 specimen has the lowest value at 35.12 N/mm². We can observe that increasing the proportion of resin in composite also increases the compressive strength of composite. The case was opposite in tensile test. There is greater variation in the result of hemp fiber composite. The variation can be reduced if the composite is prepared with consistent dimension, and conducting careful experimentation. In terms of compressive strength, hemp fiber composite possesses better strength than jute fiber. The test results, which include the compressive strength and standard deviation for each specimen, are summarized in the Table 4.

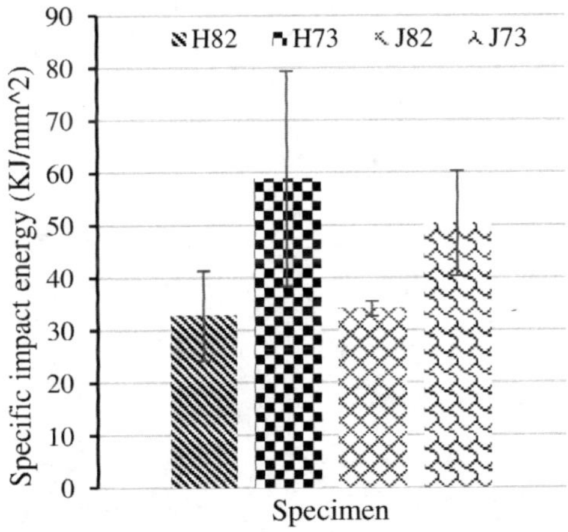

Figure 5. Specific impact energy

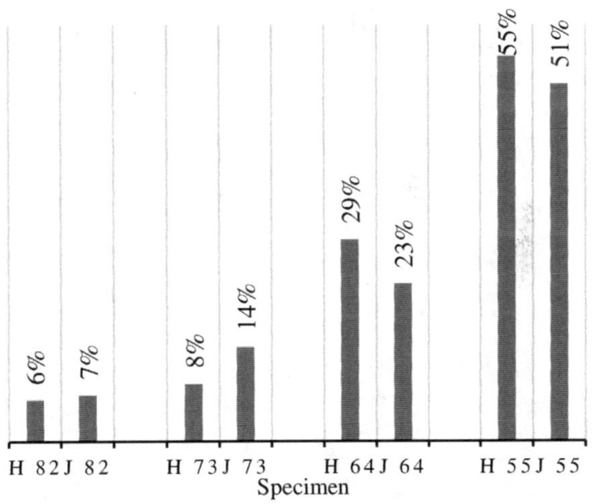

Figure 6. Water absorption % of composites

3.4. Charpy Impact Test

The Charpy impact test also reveals that hemp has better impact resistance property than jute fiber. From the result, H73 has great specific impact energy than other composite specimens. Though, there is greater variation in data. The thickness for different specimen was very difficult to maintain which slightly contributed in the variation of observed data. Though the data can be taken as reference for further experimentation of same composite in another research. For both the fiber composites, the impact energy increases with increasing fiber composition from 20% to 30% and consequently decreasing the epoxy resin composition from 80% to 70%. The result of the specific impact energy obtained from the Charpy impact test is presented in Table 5 and plotted in Figure 5.

IOP Conf. Series: Materials Science and Engineering **1314** (2024) 012005 doi:10.1088/1757-899X/1314/1/012005

3.5. Water Absorption Test of the Composites

The water absorption test was conducted to evaluate the composite material's resistance to moisture penetration and retention. Specimens were submerged in water at room temperature (24ºC) for 36 hours. The water absorption test revealed varying degrees of moisture absorption among different composites, as depicted in the bar diagram.Notably, the 80/20 hemp fiber composite (H82) exhibited the least water absorption, indicating good resistance to moisture. On the other hand, the 55% fiber composition (H55) showed the highest water absorption, suggesting that a higher proportion of fibers in the composite matrix leads to increased water absorption.

3.6. Post Test-Microscopic Analysis of the Specimens

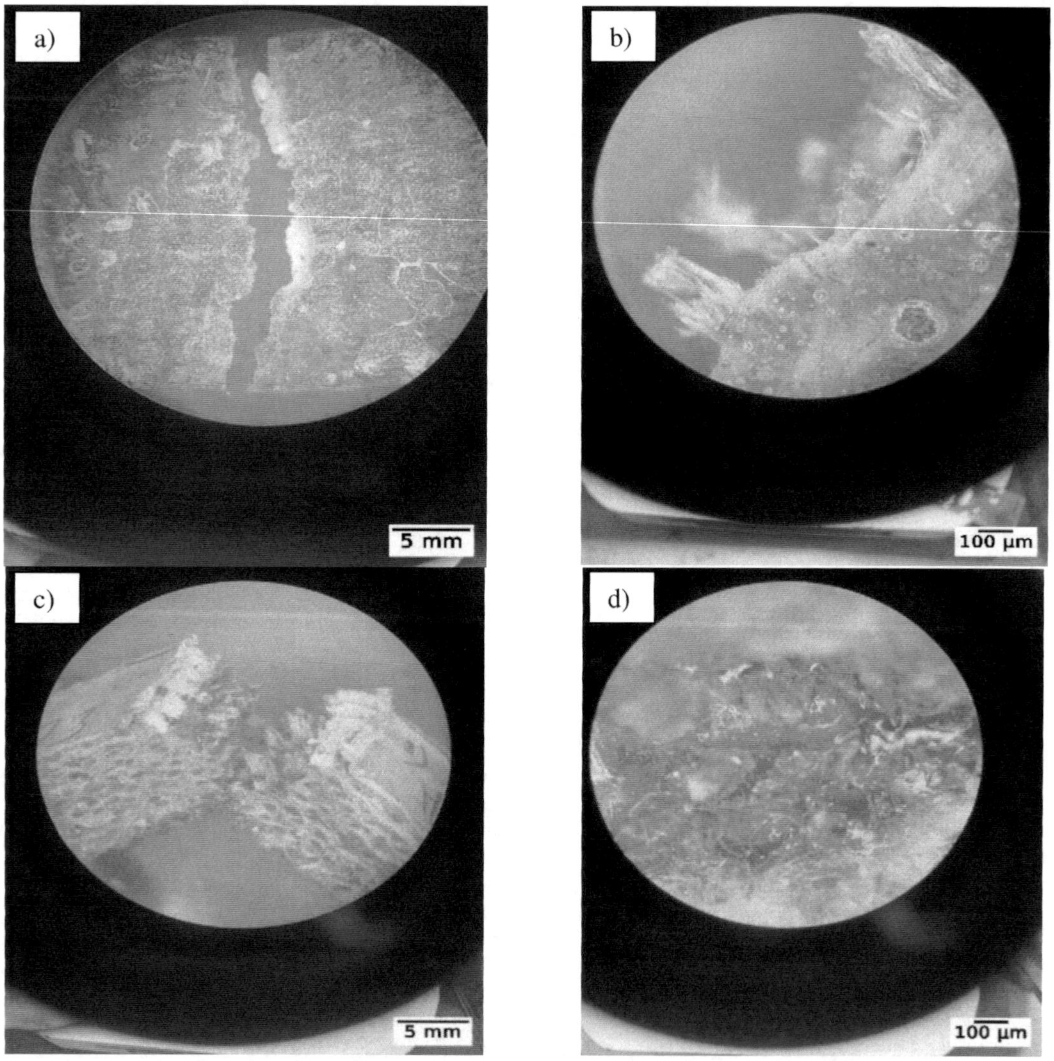

Figure 7. Failure analysis of the tensile and impact tested composites

The post-analysis was conducted using a Radical stereo zoom microscope at different magnifications to thoroughly study the test specimens after their failure under various loads. The mode of failure of specimens under tensile loading, compression loading, and impact testing is studied by observing their images. As presented in Figure 7 (a) and (b), the composite specimen under tensile load demonstrated the failure mode like pulling out of the fiber and delamination showing the improper adhesion between matrix and fiber. This failure mode affects the tensile property of the composite. This problem can be solved by using filler material which increases the adhesion between the matrix and fiber. The failure modes that occurred in Figure 7 (c) and (d) were generally due to fiber fractures, matrix cracking, and fiber pullout. The fracture of the specimen occurred at the surface of the specimen where distinctive V-shaped notch was created before the test.

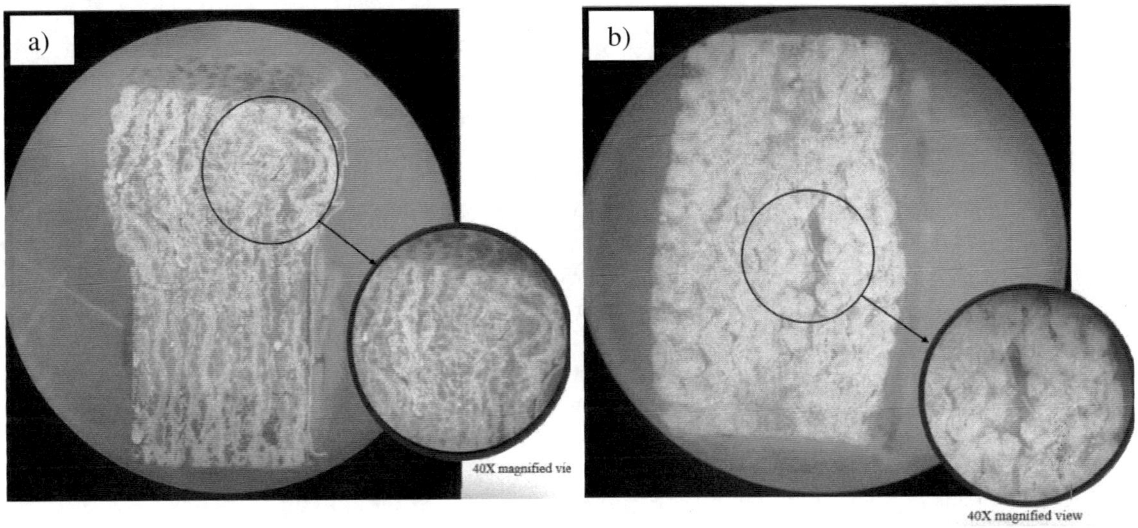

Figure 8. Failure analysis of the compression tested composites

The compression tested composite materials were studied under the microscope to understand the fracture mechanisms. The microscopic images of figure 8 (a) and (b) show the failure modes under compression load. Inter-lamina deboning occurred due to the weak interfacial bonding between the fiber and matrix or due to the geometric imperfections in the specimen, such as waviness or curvature. Figure 8 (a) shows the delamination at the top of the sample due to the bending of the fibers. These fibers were subjected to compressive stress causing them to bend and eventually delaminate. Figure 8 (b) showed a distinct delamination pattern in the center of the specimen where the deformation was concentrated. The interface was weakened due to improper bonding or weak adhesive, which caused layers to separate when the specimen was subjected to compressive load.

4. Conclusion

The study presents a comprehensive analysis of the mechanical properties of natural fiber polymer composites, with a focus on hemp and jute fibers reinforced with epoxy resin. The composites were fabricated using the hand lay-up method with varying fiber weight fractions (10% to 50%) and experimented for water absorption, tensile strength, compressive strength and Impact strength. The results observed indicates that the hemp fiber composites exhibit superior in mechanical properties to that of jute fiber composites. The hemp fiber composite with a 30% fiber weight fraction (H73) demonstrated the highest tensile strength (50 MPa), compressive strength (55.65 N/mm^2), and impact strength (58.87 kJ/m^2), while also exhibiting lower water absorption (7%) compared to the jute fiber composite (J73). This suggests that hemp fibers provide better tensile, compressive, and impact resistance and are less prone to moisture penetration. In addition, the microscopic post-analysis of test specimens after failure demonstrated similar modes of failure- delamination, fiber pull out, and buckling, indicating less interfacial bonding between fibers and the matrix. Therefore, future research should focus on enhancing the bonding between the matrix and the fiber, ensuring dimensional control during specimen preparation. Also, it's important to conduct additional tests such as fatigue and thermal analysis, and explore alternative fabrication methods like vacuum-assisted resin transfer molding

(VARTM) or compression molding under pressure to avoid forming voids and bubble form in composites, and achieve more consistent and precise results for natural fiber composites' mechanical properties.

Reference

[1] R. Hsissou, R. Seghiri, Z. Benzekri, M. Hilali, M. Rafik, and A. Elharfi, "Polymer composite materials: A comprehensive review," *Compos Struct*, vol. 262, p. 113640, Apr. 2021, doi: 10.1016/J.COMPSTRUCT.2021.113640.

[2] L. Mishnaevsky, K. Branner, H. N. Petersen, J. Beauson, M. McGugan, and B. F. Sørensen, "Materials for Wind Turbine Blades: An Overview," *Materials 2017, Vol. 10, Page 1285*, vol. 10, no. 11, p. 1285, Nov. 2017, doi: 10.3390/MA10111285.

[3] M. S. H. Al-Furjan, L. Shan, X. Shen, M. S. Zarei, M. H. Hajmohammad, and R. Kolahchi, "A review on fabrication techniques and tensile properties of glass, carbon, and Kevlar fiber reinforced rolymer composites," *Journal of Materials Research and Technology*, vol. 19, pp. 2930–2959, Jul. 2022, doi: 10.1016/J.JMRT.2022.06.008.

[4] G. R. Kalagi, R. Patil, and N. Nayak, "Experimental Study on Mechanical Properties of Natural Fiber Reinforced Polymer Composite Materials for Wind Turbine Blades," *Mater Today Proc*, vol. 5, no. 1, pp. 2588–2596, Jan. 2018, doi: 10.1016/J.MATPR.2017.11.043.

[5] Y. G. Thyavihalli Girijappa, S. Mavinkere Rangappa, J. Parameswaranpillai, and S. Siengchin, "Natural Fibers as Sustainable and Renewable Resource for Development of Eco-Friendly Composites: A Comprehensive Review," *Front Mater*, vol. 6, Sep. 2019, doi: 10.3389/FMATS.2019.00226.

[6] J. K. Pandey, S. H. Ahn, C. S. Lee, A. K. Mohanty, and M. Misra, "Recent Advances in the Application of Natural Fiber Based Composites," *Macromol Mater Eng*, vol. 295, no. 11, pp. 975–989, Nov. 2010, doi: 10.1002/MAME.201000095.

[7] E. Zini and M. Scandola, "Green composites: An overview," *Polym Compos*, vol. 32, no. 12, pp. 1905–1915, Dec. 2011, doi: 10.1002/PC.21224.

[8] G. R. Kalagi, R. Patil, and N. Nayak, "Experimental Study on Mechanical Properties of Natural Fiber Reinforced Polymer Composite Materials for Wind Turbine Blades," *Mater Today Proc*, vol. 5, no. 1, pp. 2588–2596, 2018, doi: 10.1016/J.MATPR.2017.11.043.

[9] "Assessment of Research Needs for Wind Turbine Rotor Materials Technology," *Assessment of Research Needs for Wind Turbine Rotor Materials Technology*, Jan. 1991, doi: 10.17226/1824.

[10] S. Rana and R. Fangueiro, "1 - Advanced composites in aerospace engineering," in *Advanced Composite Materials for Aerospace Engineering*, S. Rana and R. Fangueiro, Eds., Woodhead Publishing, 2016, pp. 1–15. doi: https://doi.org/10.1016/B978-0-08-100037-3.00001-8.

[11] A. Gholampour and T. Ozbakkaloglu, "A review of natural fiber composites: properties, modification and processing techniques, characterization, applications," *J Mater Sci*, vol. 55, no. 3, pp. 829–892, 2020, doi: 10.1007/s10853-019-03990-y.

[12] D. U. Shah, P. J. Schubel, and M. J. Clifford, "Can flax replace E-glass in structural composites? A small wind turbine blade case study," *Compos B Eng*, vol. 52, pp. 172–181, 2013, doi: https://doi.org/10.1016/j.compositesb.2013.04.027.

[13] J. W. Holmes, P. Brøndsted, B. F. Sørensen, Z. Jiang, Z. Sun, and X. Chen, "Development of a Bamboo-Based Composite as a Sustainable Green Material for Wind Turbine Blades," *Wind Engineering*, vol. 33, no. 2, pp. 197–210, Mar. 2009, doi: 10.1260/030952409789141053.

[14] K. Palanikumar, M. Ramesh, and K. Hemachandra Reddy, "Experimental Investigation on the Mechanical Properties of Green Hybrid Sisal and Glass Fiber Reinforced Polymer Composites," *Journal of Natural Fibers*, vol. 13, no. 3, pp. 321–331, May 2016, doi: 10.1080/15440478.2015.1029192.

[15] A. Felix Sahayaraj, M. Muthukrishnan, and M. Ramesh, "Experimental investigation on physical, mechanical, and thermal properties of jute and hemp fibers reinforced hybrid polylactic

acid composites," *Polym Compos*, vol. 43, no. 5, pp. 2854–2863, May 2022, doi: 10.1002/PC.26581.

[16] N. Saba, M. Jawaid, and M. T. H. Sultan, "1 - An overview of mechanical and physical testing of composite materials," in *Mechanical and Physical Testing of Biocomposites, Fibre-Reinforced Composites and Hybrid Composites*, M. Jawaid, M. Thariq, and N. Saba, Eds., Woodhead Publishing, 2019, pp. 1–12. doi: https://doi.org/10.1016/B978-0-08-102292-4.00001-1.

[17] M. Chowdhury and S. Hossain, "A Review on Tensile and Flexural Properties of Fiber-Reinforced Polymer Composites," pp. 16–26, Nov. 2020, doi: 10.9790/019X-07051626.

[18] S. S. Raj, K. A. Michailovich, K. Subramanian, S. Sathiamoorthyi, and K. Kandasamy, "Philosophy of Selecting ASTM Standards for Mechanical Characterization of Polymers and Polymer Composites," *Materiale Plastice*, 2021, [Online]. Available: https://api.semanticscholar.org/CorpusID:244198456

Design and Development of a Wall-Pressed Robot for In-Pipe Inspection

Raghwendra Prasad Singh[1], Prajwal Dahal[1], Safal Shrestha[1], Sirapa Shrestha[1] and Chiranjeevi Mahat[1]

[1] Department of Mechanical Engineering, Kathmandu University, Dhulikhel, Nepal

sirapa.shrestha@ku.edu.np

Abstract. Pipelines have become a major part of energy infrastructure, which requires a secure and reliable means of maintenance. It has become critical to introduce advanced robotic methods for inspecting pipelines in order to offer a more secure alternative to human entry and facilitate deployment of specialized visual inspection devices. This paper primarily deals with the design and development of a simple wall-pressed robot capable of detecting visual defects inside pipelines of diameters ranging from 38 cm to 65 cm. The overall structure of the robot is divided into two segments, each equipped with a three-wheel module arranged 120° from the central axis. Static structural analysis for strength optimization is performed using finite element method. Poly lactic acid plastic and aluminium alloy are selected as structural materials for lightweight design. Real time forces on the robot structure are simulated by applying fixed supports at the holding frame, – 33.85 N compressive forces at the wheel ends of the linkage in the axial direction and placing 145 gm weight at the motor linkage. Motion simulations are carried out to study functionality of driving mechanism. A robot prototype is built and powered by 11.1 V 5200 mAh LiPo batteries. Experimental testing is conducted in a steel pipe of inner diameter 38 cm to validate the robot's design, assessing traction, gradeability and operational robustness. Testing results are validated with calculated results that the robot applied a tractive force of 26.26 N to the in-pipe wall against each wheel during normal operation and maintained a uniform velocity of 30 mm/sec in horizontal test. This robot has the potential to ensure the safety, reliability and affordability of operations in oil and gas pipelines.

Keywords. Wall-pressed robot, in-pipe inspection, motion study, static analysis

1. Introduction

Pipelines play a vital role in the global transportation of oil, gas and various liquids [1]. Given the growing demand for these resources, the expansion of pipeline networks has become essential. However, it is important to ensure the lifespan of these pipelines, as risks such as corrosion, cracks and defects can lead to leaks and other issues [2]. Regular inspections are necessary to maintain safety and reliability. For this task,

In-Pipe Inspection Robots (IPIR) are becoming increasingly popular. These small robots are equipped with cameras and sensors to reach inaccessible areas [3] and perform inspections independently [4], improving efficiency while reducing the risk of accidents [5]. Remote-controlled IPIRs can navigate pipelines of different sizes and shapes, making them a versatile solution for pipeline maintenance and safety.

There are different types of IPIRs based on different locomotion mechanisms. Wheel-type IPIRs [6] – the most common type – use wheels powered by actuators/gear motors to move inside pipelines. Wall-pressed IPIRs [7] press against the walls of pipe with the help of spring/screw rod with linkage mechanism to propel themselves. Screw-type IPIRs use helical drives to move in a spiral way [8]. Walking IPIRs [9] have 4-6 legs that can independently control to move through complex pipe geometries. Inchworm [10] and Squirm [11] IPIRs uses friction to move by repetitive contraction and expansion of their bodies. Some inchworm uses various telescopic mechanism consisting of motor-screw for more efficient movement [12]. Caterpillar IPIRs employ tracks instead of wheels, which increases their surface area contact with the pipe [13]. PIG (Pipe Inspection Gauge) IPIRs are passive robots that are driven by the flow of fluid in pipe [14]. In addition to these basic types, a variety of hybrid IPIRs combine different mechanisms for specific applications [15]. Table 1 shows applications, advantages and disadvantages of the above-mentioned robots.

Table 1. Chart showing applications, advantages, and disadvantages of different types of robots [6]–[15].

S.N.	Types of Robots	Application	Advantage	Disadvantage
1	Wheel	Mostly used for horizontal pipelines	Very simple in design and moves in high speed	Cannot move in complex (irregular) pipe and low traction
2	Wall press	For both horizontal & vertical pipelines	High traction	Cannot move in complex pipe
3	Screw	For both horizontal & vertical pipelines	Low weight and high speed	Reversing the motion of the robot is difficult
4	Walking	For any type of surface like rough, irregular	Can avoid obstacles and low weight	Has complex structure
5	Inchworm & Squirm	Mostly for small diameter pipelines	Squeeze through tight areas	Slow speed and drive force is small
6	Caterpillar	For Supporting heavy body weight	High stability and moves over obstacles in pipeline	Slow speed and heavy weight
7	PIG	Especially for large diameter pipelines	High speed and travels long distances	Needs liquid medium to travel and suitable for one particular diameter.

2. Literature review

Recent advances in IPIRs have focused on developing robots with adaptable structures, autonomous navigation, and the ability to achieve vertical and curved motion. Hua Song et al. [12] proposed a telescopic IPIR that uses a step motor-screw driven mechanism to adjust its length. Jinwei Qiao et al. [16] also developed the telescopic robot with a combination of cylinder and spring mechanism for similar adaptability. Suzumori et al. [17] described the design of a micro in-pipe robot for 1-inch pipe that used camera for micro cracks and dual hand for manipulating small objects in pipes. Toyomi Miyagawa et al. [18] described the running static analysis deployed on a pipeline within-pipe mobile robot using a wheel drive mechanism for straight pipelines. The wheel mechanism, driven by planetary and worm gears, can both drive the wheel and press against the pipe wall using only actuators and simple mechanisms.

Bhadoriya et al. [19] presented a design of in-pipe robot having low weight which passes into large diameter. The driving wheel of the robot was driven by a battery-operated motor, while the backup wheels were driven by a DC motor through spur gears and links. The link was mounted on the gear and its other end was concentric with the center of the backup wheel. Ori Inbar et al. [20] designed reconfigurable field robot actuated with 4 motors having the sprawl mechanism, allowing the robot to change its width and the four-bar extension mechanism, enabling it to adjust its height and shift its center of mass. Dr. Sharanabasappa et al. [21] proposed an adjustable in-pipe robot having a central rod upon which a translational element was fitted, which was connected to three frames of links. It was also equipped with a variety of sensors, including cameras and ultrasonic sensors, which allowed it to autonomously navigate pipelines. Elankavi et al. [6] presented the simulation of IPIRs with wheels at different angles compared to 120° each and showed that the proposed robot provided equal force distribution in all 3 wheels. Table 2 shows the specifications of different robots used in the above mentioned literatures.

Table 2. Chart showing specifications of different robots used in literatures.

Ref.	Types of Robots	Pipe Configuration	Motor RPM Used	Diameter Adaptability (mm)	Traction Force (N)	Horizontal Velocity (mm/s)	Vertical Velocity (mm/s)
[12]	Inchworm	Horizontal and vertical	-	>70	4.2-187.8	-	-
[16]	Inchworm	Horizontal and vertical	-	19	15.2	12	9.3
[17]	Micro Planetary Gear Wheel Drive	Vertical and curved	-	23	1	-	6
[18]	Wheel Drive with Planetary Gears	Horizontal	-	48	20	11	-
[19]	Wheel with Gear	Horizontal and vertical	100	150	1.8 Horizontal and 3.8 Vertical	42.25	38.43
[20]	Wheel and Wall-Pressed Robot	Horizontal, vertical and transition	-	120-330	10-12	400	250
[21]	Wheel and Wall Pressed Robot	Horizontal and vertical	60	200-260	6.1	300	-
[6]	Wheel	Horizontal and curved	100	250-350	2.5-12	330	-

All of the robots in Table 2 were capable of operating in horizontal pipes however only some of them were capable of operating in vertical and curved pipes. Some inchworm robots used motors for gear or telescopic action, but all wheel type robots required high powered motors. The diameter adaptability of the robots varies from 19 mm to 350 mm. The robots were used in a wide range of pipe sizes. The traction force of the robots varies from 1 N to 187.8 N that the robot can apply to the pipe wall to move forward. The velocity of the robots varies from 3.8 mm/s to 400 mm/s. The velocity of the robot depended on the motor rpm used and various other specifications of robot.

Julianne et al. [22] presented a method of rust detection using image processing. Rust on an image can be identified by setting a threshold value and applying edge detection and segmentation techniques. Figure 1 shows the detection of rusting which can be observed in pipes during inspection. Using this type of procedure, cracks, leaks, and even holes can be detected.

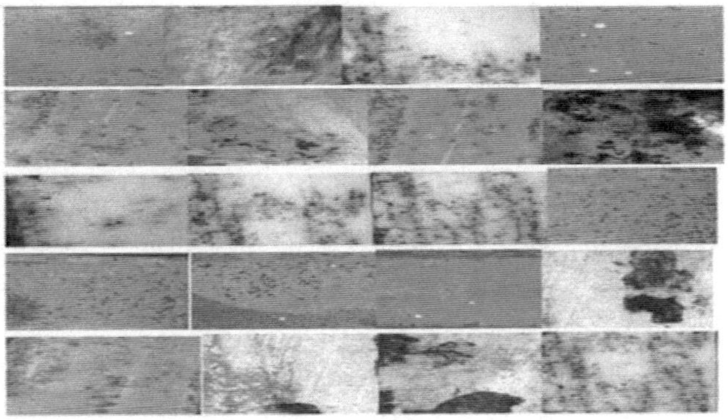

Figure 1. Use of image processing to detect rust clearly[22]

Mingzhu et al. [23] shows a framework for tracking multiple sewer defects in CCTV videos by computer vision techniques. First, a deep learning-based defect detection model and a metric learning model were developed and trained using the sewer datasets. Then, the tracking module uses the results and features of these trained models to predict traces using a Kalman filter and associate these traces based on defect motion, appearance characteristics, and defect type. Figure 2 displays an actual camera image of the detection process. This real-time defect identification is achieved by comparing the camera image to the reference images using image processing and machine learning techniques.

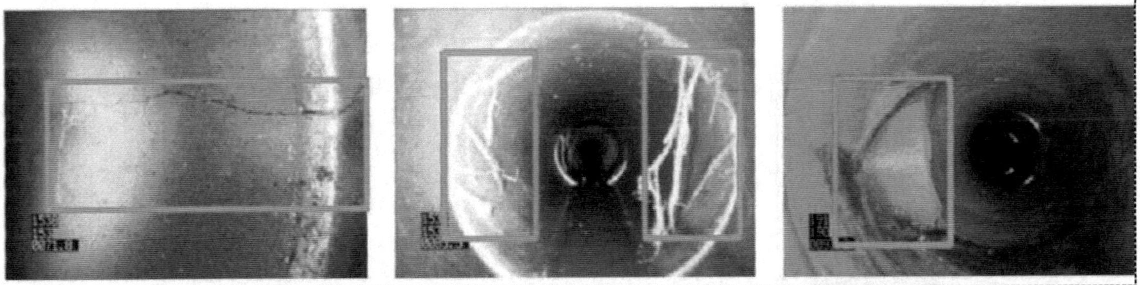

Figure 2. Use of computer vision techniques for real time defect detection[23]

3. Materials and methods

The robot is divided into two segments, each equipped with a three-wheel module arranged 120° from the central axis. The robot is fitted with six slider crank mechanism [24] which allows to expand with diameters ranging from 38 cm to 65 cm with sufficient traction force to the wheel. Analytical calculations are done to find robot's functional parameters during operating conditions and force analysis is conducted. CAD modelling is performed using SolidWorks and material selection is done using Ashby Methodology. For checking whether the robot design is safe or not, analysis is done in ANSYS. The robot's motion inside a pipe is simulated in SolidWorks to analyze its velocity and reaction force over time. This is followed by an

IOP Conf. Series: Materials Science and Engineering **1314** (2024) 012006 doi:10.1088/1757-899X/1314/1/012006

experimental test in a horizontal pipe to validate the simulation results. Figure 3 shows the process flow diagram of the research:

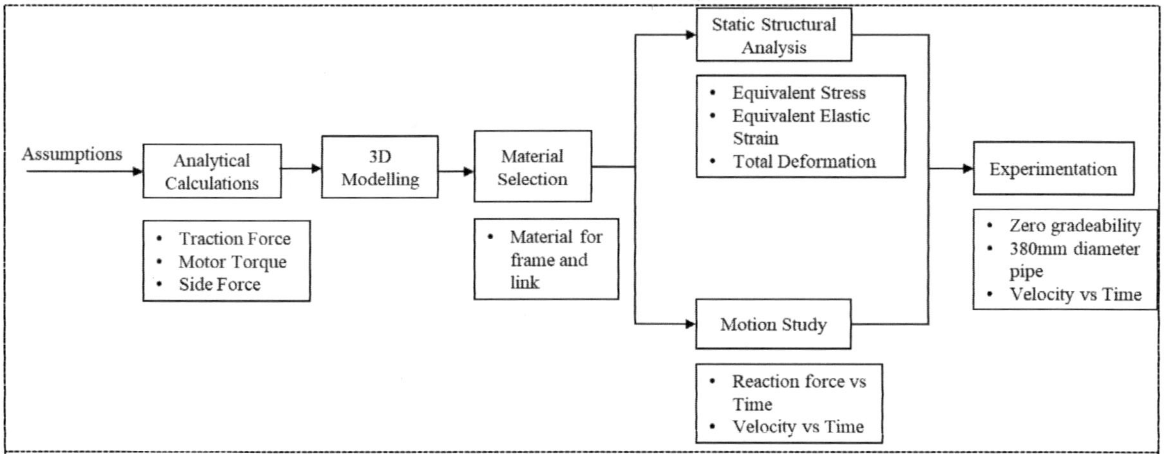

Figure 3. Process flow diagram.

3.1. Design and calculations

Table 3 shows parameters used in the calculation.

Table 3. Parameters.

S.N.	Assumption	Values	Units	Remarks
1	Total Mass	4.05	kg	Weight of Prototype
2	Tire diameter (Refer Figure 2)	9.5	cm	Diameter of Prototype
3	Length of link-1 (OB)	23	cm	Assumed
4	Distance between screw rod and link-1 (AB)	4.44	cm	Assumed
5	Length of link-2 (CE)	14.8	cm	Assumed
6	Coefficient of friction between steel pipe and tire (μ)	0.3-0.6		Assumed
7	Acceleration due to gravity (g)	9.81	m/s^2	Universal constant
8	Angle between axis BH and link 1 (θ)	60^0		Assumed
9	Number of revolutions	10	rpm	Motor power of prototype

Straight horizontal and vertical configurations of pipe is selected based on the application of robot to be employed. To traverse successfully through these configurations of pipes, the kinematic scheme of the robot is finalized. A kinematic scheme helps to analyse robot motion, planning, control and analysis of its movement. This is achieved by deriving constraints equations for the design of the robot module. To represent 3D design in simple way, a line diagram is drawn which is useful to derive the constraint equations of diameter of pipe (refer Table 4 step 2) and length of the robot (refer Table 4 step 3). In the Figure 4, R is radius of the pipe, L is the length of the larger link OB, AB denotes the distance between central axis of screw rod and larger link OB, OD is the radius of the driving wheel, θ is the included angle between link OB and axis BH, β is the included angle between link OB and CG.

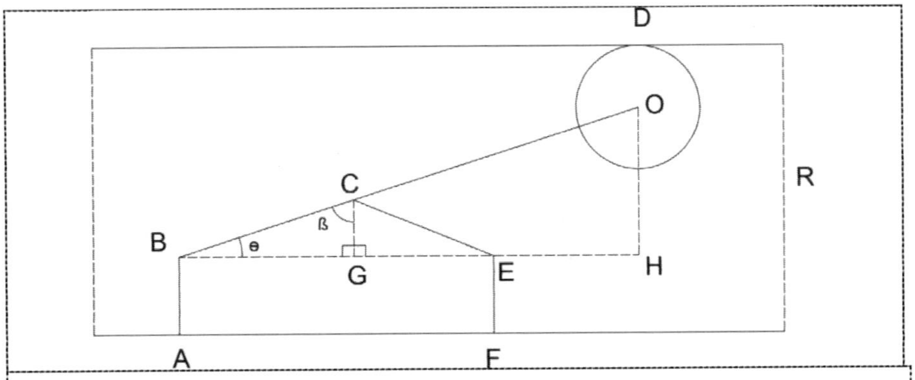

Figure 4. Line diagram of robot.

Depending on the robot driving condition, the robot wheel needs to apply side force for grip. The robot must be able to overcome this side force to generate enough traction force to move forward. In the Figure 5, for horizontal condition, N1, N2 and N3 represents the three sets of normal force acting on each wheel and $\mu_s \cdot N$ denotes the frictional force acting in tangential direction. A weight mg is acting vertically downward from the central axis of the robot. Likewise in Figure 6 for vertical climbing, the robot needs to apply enough propelling force to overcome self-weight and gravity. As the design is symmetric, considering the friction and side force equally divided between upper and lower wheel module, the average friction force per wheel is one-sixth of mg.

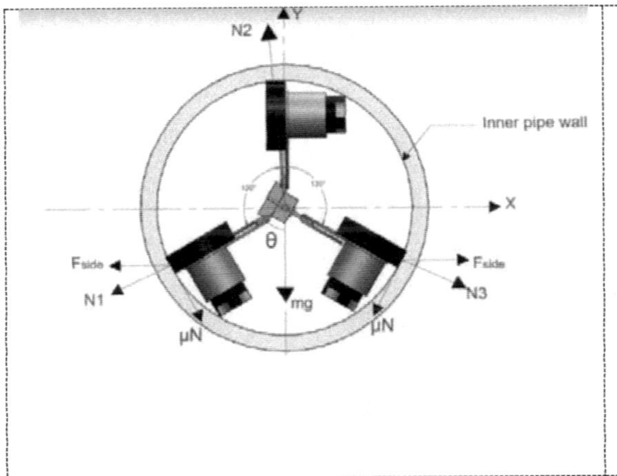

Figure 5. Free Body Force Analysis diagram for horizontal configuration.

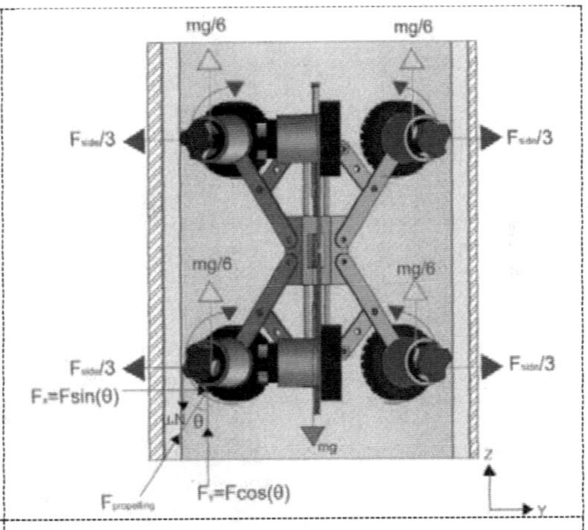

Figure 6. Free Body Force Analysis diagram for vertical configuration.

The equations related to force analysis are tabulated below in Table 4.

Table 4. Equations used for mechanical design and parameter selection.

S.N.	Parameters	Equation	Value
1	OH	$OB \sin\theta$	14.09 cm
2	Total diameter of pipe **(D)**	$2(\mathrm{AB} + \mathrm{OH} + \mathrm{OD})$	38 cm

3	Length of Robot (***AF***) Horizontal Driving	$BC\ Cos\ \theta\ +\ CE\ Cos\ \theta$	$50\ cm$
4	Traction force (***F***)	$\frac{mg\cos\theta\cos\gamma}{k_1}(k_2k_4\mu + k_3)$	$26.26\ N$
5	Torque Required (***T***)	$\text{F} \times \text{r}$	$1.24\ Nm$
6	Normal reaction (***N***)	$\frac{mg}{cos\theta - \mu sin\theta}$	$33.85\ N$
7	Side Force ($\boldsymbol{F_{side}}$)	$\frac{mg(sin\theta + \mu cos\theta)}{cos\theta - \mu sin\theta}$	$36.085\ N$
8	Vertical Climbing Propelling Force	$\frac{mg}{cos\theta - \mu sin\theta}$	$58.8\ N$
9	Torque on each wheel	$\frac{F\ r}{3}$	$0.93\ N$
10	Power (***P***)	$T * \frac{2\pi n}{60}$	$3.3\ W$

Here, k1, k2, k3, and k4 are supporting forces whose values are calculated by using formulae from Y. Zhang et al. [25] in step (4).

3.2. CAD modelling

CAD modelling was performed using SolidWorks shown in Figure 7. Individual components are designed on the basis of theoretical calculations. Then the components are assembled together, and a 3D design is prepared.

Figure 7. CAD model of robot.

3.3. Material selection

Ashby methodology is employed for the material selection. The following four steps are involved in Ashby methodology: Identifying desirable characteristics, Screening, Ranking and Documentation and

Localization [26]. Bending resistance, light weight holding frame, material availability and low cost were taken into consideration to choose the best material that meets the needs of the design. The final material selection is shown in Table 5.

Table 5. Material selection using Ashby methodology.

S.N.	Component Name	Material Property	Material Selected	Quantity
1	Holding Frame	High strength, lightweight & corrosion resistant	PLA Plastic and Mild steel	1
2	Screw Rod	High strength & corrosion resistant	Stainless steel	1
3	Nut	High strength & corrosion resistant	PLA Plastic	2
4	Wheel	Flexible	Rubber	6
5	Check nut	High strength & corrosion resistant	Stainless steel	12
6	Connecting link (larger and smaller)	High strength to weight ratio	Aluminum Alloy	12

3.4. Static structural analysis

The static structural analysis has been performed to analyze the structural strength of the mechanism against external loading conditions like compressive force exerted by inner pipe wall. This analysis involves several steps. First, the mesh is generated and refined until the results converge. Various materials, such as aluminum alloy (for small and large connecting link), low carbon steel (for screw rod), and PLA plastic (for holding frame), were tested to evaluate component strength. Real time forces on the robot structure were simulated by applying fixed supports at the two ends of holding frame, 33.85 N compressive forces at the wheel ends of the linkage in the axial direction and placing 145 gm combined weight of motor and tyre at the motor linkage. After the analysis setup, the simulation is run and the results are analyzed. Figure 8 represents the boundary condition.

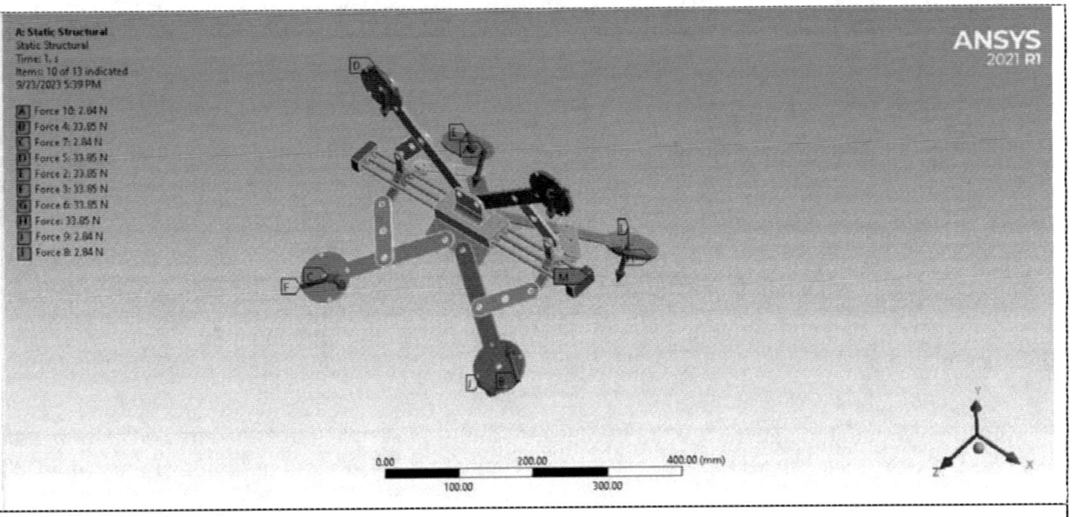

Figure 8. Boundary condition showing fixed support and force applied.

3.5. Motion study

After completing 3D design and static structural analysis, motion study for robot's horizontal and vertical configuration in pipeline is performed in SOLIDWORKS. Motion analysis of the inspection robot inside the pipe is performed by assembling the robot into the pipe and considering various parameters like 9.8 m/s^2 gravity, 0.4 coefficient of friction and 10 rpm of motor speed in all six wheels (for horizontal configuration). The length and diameter of pipe is set to be 4 m and 0.45 m respectively. Velocity vs Time and Reaction force vs Time graphs are plotted. Simulations for vertical and inclined conditions are performed as well with minor changes which will be addressed in result section. Figure 9 represents motion analysis of robot.

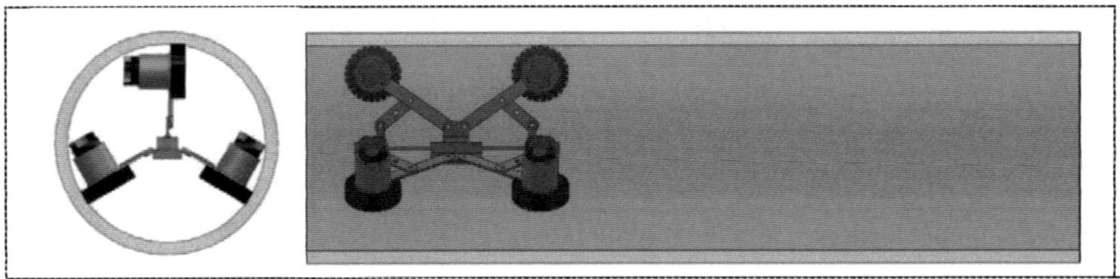

Figure 9. Robot motion analysis.

3.6. Experimental setup

For the experiment, a steel pipe of diameter 0.38 m and length 4 m is selected. It is kept horizontally (maintaining 0°) with the ground. The robot is fitted with a webcam (ESP-32) that is able to capture visual defects in the pipeline. The width of the robot is adjusted at 0.38 m from 0.45 m in order to fit inside the pipe. For propulsion, the robot is fitted with six DC motors of 10 rpm. The robot prototype is powered by 11.1 V 5200 mAh LiPo batteries and is connected by wires. Then the robot is placed on one end of the pipe and is allowed to move continuously to the other end. Time required to travel each 0.4 m of the 4 m pipe is recorded using a stopwatch to calculate the velocity, and corresponding velocity versus time graph is plotted. Figure 10 represents the experimental setup for horizontal configuration of pipe.

Figure 10. Experimental setup for horizontal configuration.

4. Result and discussion

4.1. Static structural analysis

The simulation result is compared with the calculations. Equivalent stress, equivalent strain and total displacement are the major parameters considered to evaluate the structural behavior of the robot under the applied conditions. Figure 11 represents the maximum equivalent stress (τ_{design}) of 36.603 N/ mm^2 while the minimum equivalent stress of 0 N/mm^2 indicates the areas with the least amount of displacement. Figure

12 represents the maximum equivalent elastic strain of 0.00035372 N/ mm^2 while the minimum deformation of 0 N/mm^2 indicates the areas with the least amount of displacement. Figure 13 represents the maximum deformation of 0.39755 N/ mm^2 while the minimum deformation of 0 N/mm^2 indicates the areas with the least amount of displacement. The calculated allowable shear stress for aluminum alloy link [27] is:

$$\tau_{allow} = \frac{Maximum\ Shear\ Stress}{Factor\ of\ Safety} = \frac{207}{2} = 103.5\ N/mm^2$$

Since, Design Shear Stress (τ_{design}) = 36.603 N/mm^2

$\because \tau_{design} < \tau_{allow}$, the design is safe.

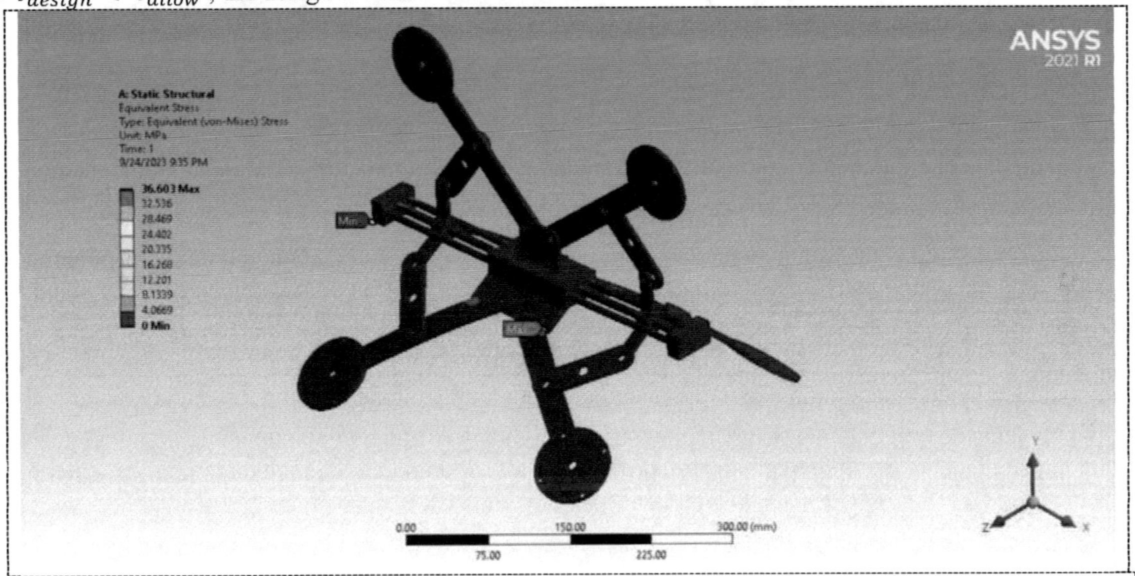

Figure 11. Equivalent stress analysis.

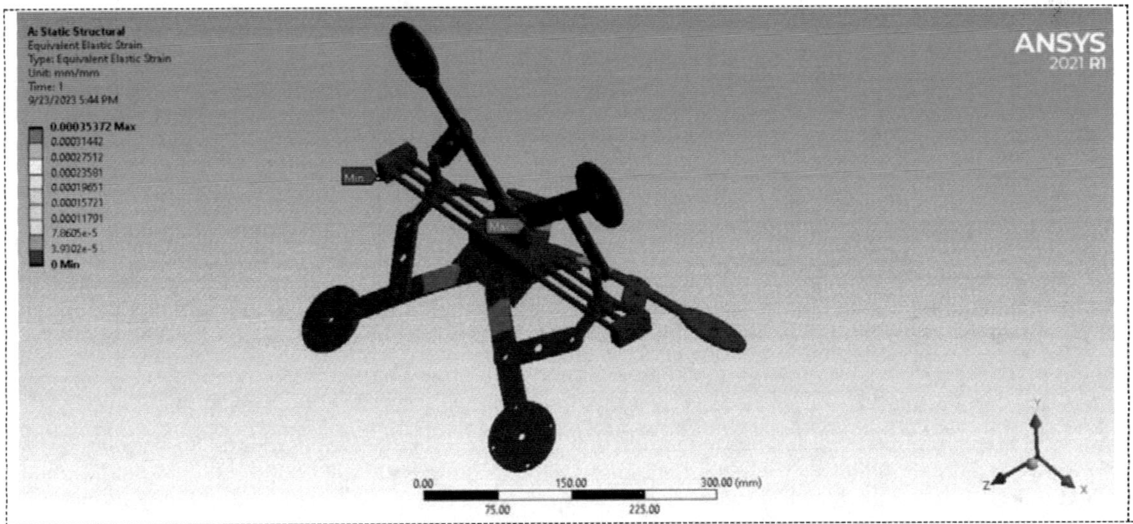

Figure 12. Equivalent elastic strain.

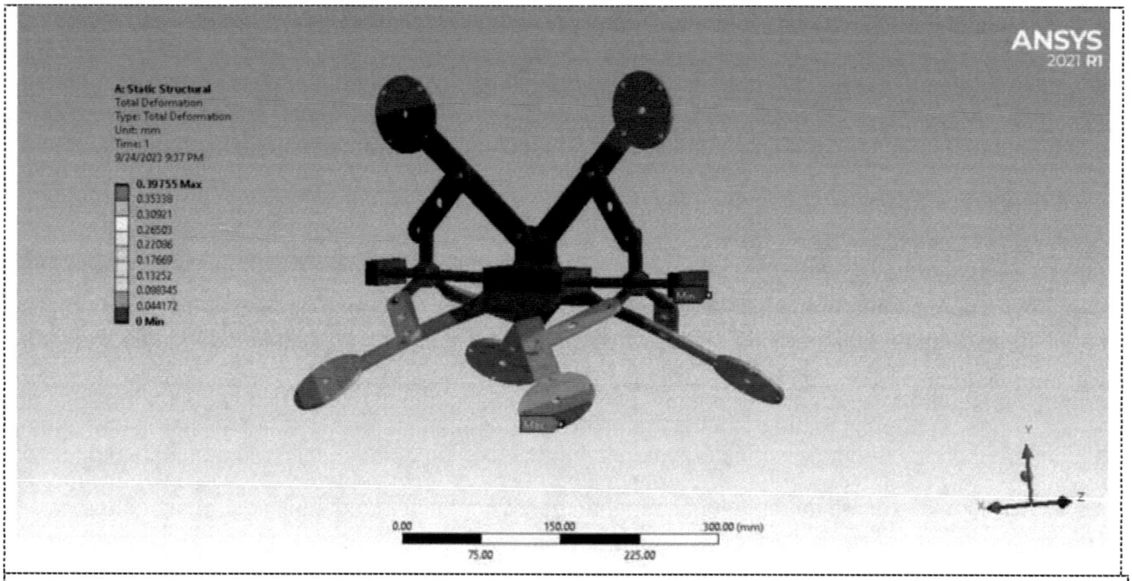

Figure 13. Total deformation.

4.2. Motion study

Figure 14 shows the linear velocity of the robot when travelling inside the pipeline. A constant velocity of 56.05 mm/s is achieved within 0.07 sec. The instantaneous rise in velocity is to overcome the traction force by wheels of the robot. Experimental result of velocity Vs time is given in figure 15.

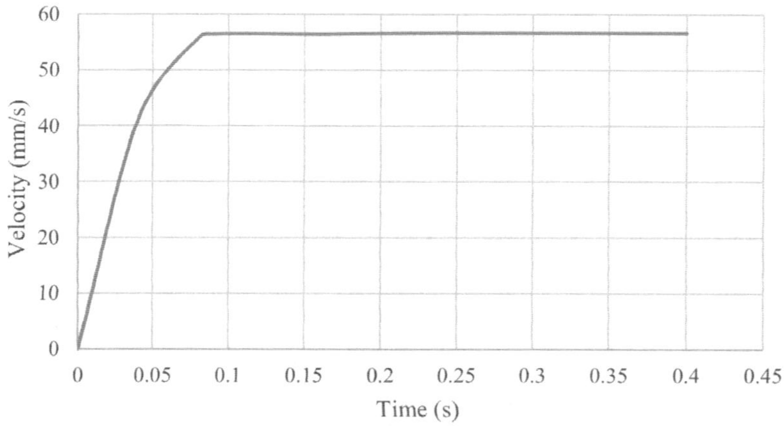

Figure 14. Variation of velocity with time for 3D model robot moving in a horizontal pipe.

Figure 15 shows the reaction force in simulation ranging from 23.37 N to 33.21 N. It is observed that the reaction force from theoretical calculation i.e. 33.85 N which is slightly larger than the simulated average of 29 N. This is because the dynamic effect considered for reaction force calculation is not same in both the case. The robot moves continuously during simulation study whereas static theoretical calculation lacks this effect.

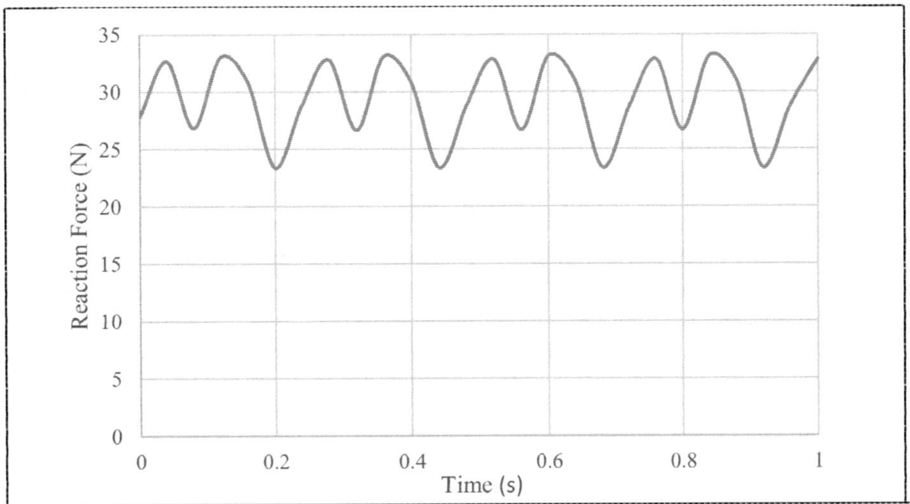

Figure 15. Variation of reaction force with time.

Elankavi et al. [6] describes the forces on the wheels of an in-pipe robot, which vary depending on the pipe diameter and curvature as shown in Figure 15. The pattern in Figure 15 shows the spike of reaction force with time which is close to pattern in Figure 16. The value of different reaction force is due to 100 rpm of motor by Elasnkavi et al. Also, the rise and fall of reaction force is due to frequent contact and no contact between wheels and pipelines. As the literature suggested, the robot's links should be at different angles rather than mounting at 1200 for even force distribution.

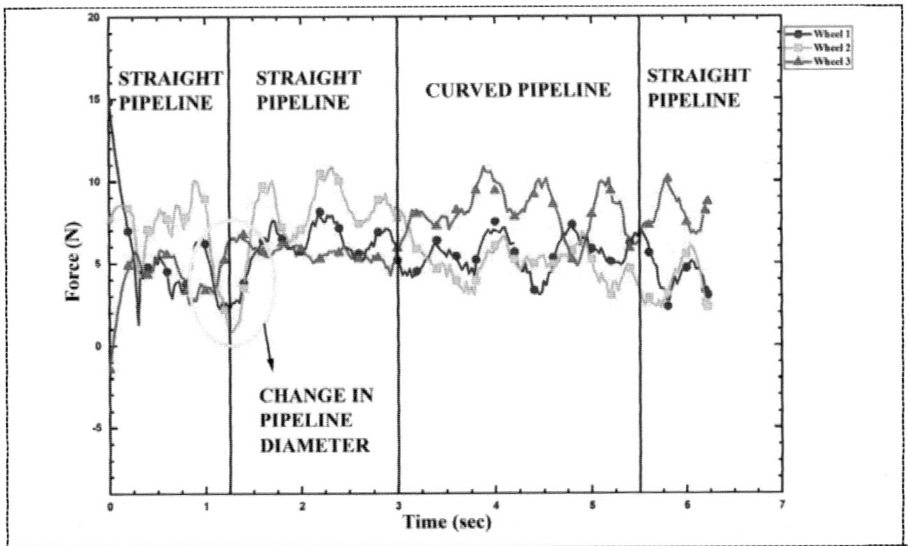

Figure 16. Force acting on wheels of the robot when travelling in straight, curved and varying diameter pipeline as per Elankavi et al. [6].

The robot can only overcome a 7.5^0 inclination with 10 rpm of the motor. When the inclination is increased beyond that, the robot fell down. Therefore, the rpm of the motor needs to be at least 75 rpm to run in a vertical pipeline based on simulation. Based on Table 2 in literature review, motors with rpms of 60-100 were used for vertical climbs. Since the robot's motor is only an rpm of 10, it can only be applicable for inclinations of 0^0 to 7.5^0. Other optimizations can be done to improve the performance of the robot, such as weight reduction, mechanism change, and using a high torque motor.

4.3. Experimental result

The experiment is conducted in a 4 m pipe, the velocity seems to be constant after few seconds during start. So, velocity graph as shown in Figure 17 is taken for short interval of time highlighting the main pattern. The constant velocity was achieved in 0.22 sec. But the graph is not perfectly linear due to the obstacles the robot has to overcome during real testing conditions. The deflection of 0.15 sec is seen as compared to the simulation result. This change may be due to the loss in efficiency of the motor, coefficient of friction and other unknown factors during real working conditions. The experiment result validates the simulation very closely.

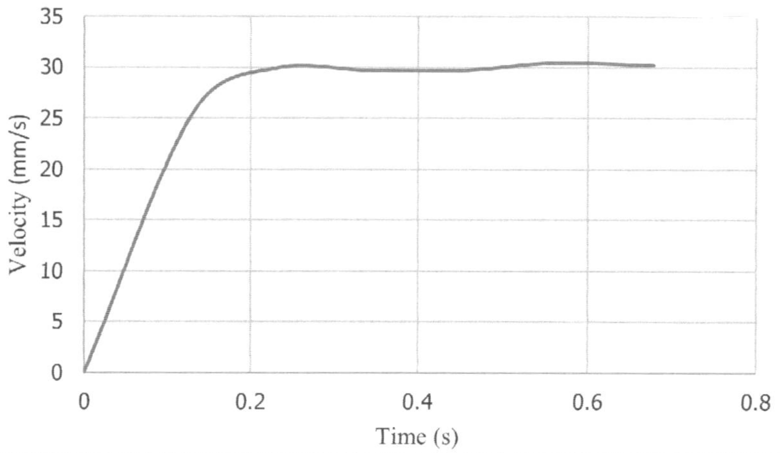

Figure 17. Variation of velocity with time for experimental robot moving in a horizontal pipe.

5. Conclusion

In this study, a simple wall pressed in-pipe robot is designed and developed capable of inspecting pipelines of diameter ranging from 38 to 65 cm. A six-slider crank mechanism is introduced to apply sufficient grip to inner pipe wall. The robot is simple, lightweight, and effective and has the potential to revolutionize the way pipelines are inspected and make them safer and more reliable. Static structural analysis is conducted to ensure the safety and reliability of design. Motion simulations and experimental testing are conducted and velocity versus time graph is validated for horizontal pipe configuration. Visual defects in pipelines are analyzed in real time using ESP-32 camera. Experimental testing in a steel pipe of inner diameter 38 cm shows that the robot can apply a tractive force of 26.26 N to the in-pipe wall against each wheel and tries to maintain a uniform velocity of 30 mm/sec in horizontal test.

Conducting experimental testing in vertical pipe configurations remains outside the scope of this research. Further improvement in motor and electronic components can ensure successful vertical movement. Enhancing the camera system with high-quality wireless cameras with 360^0 rotation, proper lighting, image processing, and adjustable mechanisms can expand coverage and enhance defect detection, especially in bent and irregular pipes making it prospects for future research.

6. References

[1] B. John and M. Shafeek, "Pipe inspection robots: a review," *IOP Conf. Ser. Mater. Sci. Eng.*, vol. 1272, no. 1, p. 012016, 2022, doi: 10.1088/1757-899x/1272/1/012016.

[2] W. Zhao, L. Zhang, and J. Kim, "Design and analysis of independently adjustable large in-pipe robot for long-distance pipeline," *Appl. Sci.*, vol. 10, no. 10, 2020, doi: 10.3390/app10103637.

[3] M. O. TĂtar and A. Pop, "Development of an in pipe inspection minirobot," *IOP Conf. Ser. Mater. Sci. Eng.*, vol. 147, no. 1, 2016, doi: 10.1088/1757-899X/147/1/012088.

[4] A. Gargade and S. Ohol, "Design and Development of In-pipe Inspection Robot for Various Pipe Sizes," *IOP Conf. Ser. Mater. Sci. Eng.*, vol. 1012, no. 1, p. 012001, 2021, doi: 10.1088/1757-899x/1012/1/012001.

[5] C. H. Karthik and P. Sreedharan, "Design and development of pipe-inspection robot with vision 360o," *J. Phys. Conf. Ser.*, vol. 2062, no. 1, 2021, doi: 10.1088/1742-6596/2062/1/012015.

[6] R. S. Elankavi, D. Dinakaran, A. S. A. Doss, R. M. Kuppan Chetty, and M. M. Ramya, "Design and Motion Planning of a Wheeled Type Pipeline Inspection Robot," *J. Robot. Control*, vol. 3, no. 4, pp. 415–430, 2022, doi: 10.18196/jrc.v3i4.14742.

[7] M. Muramatsu, N. Namiki, R. Koyama, and Y. Suga, "Autonomous mobile robot in pipe for piping operations," *IEEE Int. Conf. Intell. Robot. Syst.*, vol. 3, pp. 2166–2171, 2000, doi: 10.1109/iros.2000.895291.

[8] P. Li, S. Ma, B. Li, and Y. Wang, "Development of an adaptive mobile robot for in-pipe inspection task," *Proc. 2007 IEEE Int. Conf. Mechatronics Autom. ICMA 2007*, pp. 3622–3627, 2007, doi: 10.1109/ICMA.2007.4304148.

[9] W. Neubauer, "Spider-like robot that climbs vertically in ducts or pipes," *IEEE/RSJ/GI Int. Conf. Intell. Robot. Syst.*, vol. 2, pp. 1178–1185, 1994, doi: 10.1109/iros.1994.407465.

[10] A. Brunete, M. Hernando, E. Gambao, and A. Helicoidal, "Drive Modules for Low Diameter Pipe Inspection Multi-configurable Micro-robots *," no. May, pp. 4351–4353, 2006.

[11] Z. Yanheng, Z. Mingwei, S. U. N. Hanxu, and J. I. A. Qingxuan, "Design and Motion Analysis of a Flexible Squirm Pipe Robot," no. 50905019, pp. 527–531, 2010, doi: 10.1109/ISDEA.2010.437.

[12] H. Song, K. Ge, D. Qu, H. Wu, and J. Yang, "Design of in-pipe robot based on inertial positioning and visual detection," *Adv. Mech. Eng.*, vol. 8, no. 9, pp. 1–22, 2016, doi: 10.1177/1687814016667679.

[13] J. H. Kim, G. Sharma, and S. S. Iyengar, "FAMPER: A fully autonomous mobile robot for pipeline exploration," *Proc. IEEE Int. Conf. Ind. Technol.*, pp. 517–523, 2010, doi: 10.1109/ICIT.2010.5472748.

[14] J. Okamoto, J. C. Adamowski, M. S. G. Tsuzuki, F. Buiochi, and C. S. Camerini, "Autonomous system for oil pipelines inspection," *Mechatronics*, vol. 9, no. 7, pp. 731–743, 1999, doi: 10.1016/s0957-4158(99)00031-8.

[15] N. S. Roslin, A. Anuar, M. Fairuz, and A. Jalal, "A Review : Hybrid Locomotion of In-pipe Inspection Robot," vol. 41, pp. 1456–1462, 2012, doi: 10.1016/j.proeng.2012.07.335.

[16] J. Qiao, J. Shang, and A. Goldenberg, "Development of inchworm in-pipe robot based on self-locking mechanism," *IEEE/ASME Trans. Mechatronics*, vol. 18, no. 2, pp. 799–806, 2013, doi: 10.1109/TMECH.2012.2184294.

[17] K. Suzumori, T. Miyagawa, M. Kimura, and Y. Hasegawa, "Micro inspection robot for 1-in pipes," *IEEE/ASME Trans. Mechatronics*, vol. 4, no. 3, pp. 286–292, 1999, doi: 10.1109/3516.789686.

[18] T. Miyagawa and N. Iwatsuki, "Characteristics of in-pipe mobile robot with wheel drive mechanism using planetary gears," *Proc. 2007 IEEE Int. Conf. Mechatronics Autom. ICMA 2007*, pp. 3646–3651, 2007, doi: 10.1109/ICMA.2007.4304152.

[19] A. Vikram Singh Bhadoriya, V. Kumar Gupta, and S. Mukherjee, "Development of In-pipe Inspection Robot," *Mater. Today Proc.*, vol. 5, no. 9, pp. 20769–20776, 2018, doi: 10.1016/j.matpr.2018.06.406.

[20] O. Inbar and D. Zarrouk, "Analysis of climbing in circular and rectangular pipes with a reconfigurable sprawling robot," *Mech. Mach. Theory*, vol. 173, no. January, p. 104832, 2022, doi: 10.1016/j.mechmachtheory.2022.104832.

[21] A. J. Moshayedi, S. S. Fard, L. Liao, and S. A. Eftekhari, "Design and Development of Pipe Inspection Robot Meant for Resizable Pipe Lines," *Int. J. Robot. Control*, vol. 2, no. 1, p. 25, 2019, doi: 10.5430/ijrc.v2n1p25.

[22] J. A. I. Diaz, M. I. Ligeralde, J. A. C. Jose, and A. A. Bandala, "Rust detection using image processing via Matlab," *IEEE Reg. 10 Annu. Int. Conf. Proceedings/TENCON*, vol. 2017-Decem, pp. 1327–1331, 2017, doi: 10.1109/TENCON.2017.8228063.

[23] M. Wang, S. S. Kumar, and J. C. P. Cheng, "Automated sewer pipe defect tracking in CCTV videos based on defect detection and metric learning," *Autom. Constr.*, vol. 121, no. August 2020, p. 103438, 2021, doi: 10.1016/j.autcon.2020.103438.

[24] H. R. Choi and S. M. Ryew, "Robotic system with active steering capability for internal inspection of urban gas pipelines," *Mechatronics*, vol. 12, no. 5, pp. 713–736, 2002, doi: 10.1016/S0957-4158(01)00022-8.

[25] Y. Zhang and G. Yan, "In-pipe inspection robot with active pipe-diameter adaptability and automatic tractive force adjusting," *Mech. Mach. Theory*, vol. 42, no. 12, pp. 1618–1631, 2007, doi: 10.1016/j.mechmachtheory.2006.12.004.

[26] H. R. Shercliff and M. F. Ashby, "Elastic Structures in Design," *Ref. Modul. Mater. Sci. Mater. Eng.*, no. June 2015, pp. 1–8, 2016, doi: 10.1016/b978-0-12-803581-8.02944-1.

[27] K. B. R. K. Mahadevan, *Design Data Handbook for Mechanical Engineering in SI and Metric Units*.

Acknowledgment

The authors are grateful to EnergizeNepal Program and Project 22-05 Laboratory, Kathmandu University.

Credibility Evaluation of Sand-Casting Technique for Manufacturing Francis Runner

Ravi Poudel[1,2,4], Sailesh Chitrakar[1,2,4], Pratisthit Lal Shrestha[3], Tejesh Man Shakya[3], Zhongdong Qian[1,4], Zhiwei Guo[1,4], and Bhola Thapa[2,3,4]

[1] State Key Laboratory of Water Resources Engineering and Management, Wuhan University, China
[2] Turbine Testing Lab, Kathmandu University, Nepal
[3] Design Lab, Kathmandu University, Nepal
[4] Hydropower Research Centre for Himalayan Region

E-mail: sailesh@ku.edu.np

Abstract. Francis turbine with its intricate design and the critical nature of its components, involves several complexities and challenges for manufacturing. Francis runners are usually fabricated using high-quality alloy steel. In this study, a 2 kW model Francis runner was manufactured from brass material with sand casting. 3D printing technology was used to create separate molds for the hub, shroud, and runner blade of the Francis turbine. Each component was then sand cast and later joined by brazing. Compared to traditional casting techniques this approach proved to be faster and more cost-effective. Moreover, the defects encountered in the turbine components achieved through the method employed were observed to be less compared to that of investment casting used in previous models. The accuracy of the fabricated model was checked by conducting 3D scanning using the highly precise EinScan HX 3D scanner with a precision of 0.1 mm. The comparisons were made between the 3D scan and the original design. It was found that the deviations on the fabricated runner dimensions as compared to the original designs were below 2 mm. The efficiency measurement of the runner was made at a low head of 4.3 meters and the maximum efficiency obtained was 51.5% at 35% opening of guide vanes. Overall, this method of fabrication proved to be effective in terms of cost and time required for fabrication. This method could be a valuable alternative for smaller-scale turbine production. Further evaluation will be necessary for full-scale commercial applications of this method using high-quality alloy steel.

1. Introduction

Sophisticated machinery and skilled manpower are required for the complex process of manufacturing turbines. Hydraulic turbines are generally constructed with steel whereas few low-head small turbines have been fabricated with cast iron and bronze [1]. Companies like Nepal Hydro and Electric Limited (NHE) in Nepal specialize in repairing turbines. The maximum capacity of such companies in Nepal is limited to the repair of runners with diameters up to 4.5 meters [2]. Several other companies in Nepal are involved in the micro hydro business with the capacity of manufacturing cross-flow and Pelton turbines up to 160 kW. Fabricators in some cases cast Pelton runner buckets in India and perform machining in Nepal. In countries with great resources, hydraulic turbines are fabricated using 5-axis or 7-axis CNC machining. Utilization of these modern industrial techniques is still to be adopted in Nepal. Traditional manufacturers have been using brass and bronze casting to produce Francis turbines, but are unable to compete in the commercial industry [3]. The complexity of the shapes of the components and the

IOP Conf. Series: Materials Science and Engineering **1314** (2024) 012007
doi:10.1088/1757-899X/1314/1/012007

assembly have posed challenges to the fabrication of Francis Turbines at local levels. According to a study, there is a market for companies that could design and manufacture turbines for hydroelectric power companies. Manufacturing sectors that are already engaged in the production or renovation of mechanical components for hydropower could play a crucial role in bridging this gap of turbine suppliers. According to a preliminary study, there are more than 75% of the initial investments in this sector and sales of turbines with capacity up to 37 MW. [4]. The first attempt at fabricating turbines in Nepal began

(a) 92 kW Francis runner manufactured at NHE, Nepal (2011) [5]

(b) 2 kW Francis runner manufactured by TTL (2012) [6]

(c) 14 kW Francis runner manufactured by TTL (2018) [7]

(d) 14 kW Francis runner manufactured by TTL (2019) [7]

Figure 1. Turbine manufactured by TTL in collaboration with different Nepali Manufacturers from 2011 to 2019

with the first propeller turbine from Balaju Yantra Shala in 1962 A.D. and then followed by the production of the first cross-flow turbine in 1973 and the first Pelton turbine in 1975. In the current scenario, there are only a few companies in Nepal engaged in turbine manufacturing [8]. The past ten years of experience of the Turbine Testing Lab has made it competent in designing Francis turbines specifically tailored for operation in Nepal's river systems. The bottleneck for manufacturing turbines in Nepal has been the industry's manufacturing capabilities [9]. Although it has been more than 57 years since the first Francis turbine was installed in Nepal, no manufacturer has been able to produce even a single commercial Francis turbine. However, Nepalese manufacturers are equipped with the skills and tools necessary to produce Francis turbines up to 10 MW unit capacity. In addition to this, the decades of experience with the repair and maintenance of turbines damaged by silt erosion have enhanced the expertise of the manufacturers [7]. Several attempts to develop the competence of manufacturing Francis turbines at local levels have been made by the Turbine Testing Lab (TTL) and collaborating manufacturers. Figure

1 shows the fabricated Francis runners developed by TTL in conjunction with Nepali manufacturers from 2011 to 2019.

Francis turbines are generally used for heads ranging from 10 m to 650 m [10]. A sectional view of the Francis turbine is presented in Figure 2 showing the hub, shroud, and blades. This research focuses on the manufacturing of the small-scale model Francis runner (2 kW) capacity by method of sand casting. The existing design of the model Francis turbine available at the Turbine Testing Lab was used in this study. The different components like the hub, shroud, and blades were fabricated separately to achieve better results. This study was conducted in an attempt to develop competence in the fabrication of the Francis turbine at the local level with local sand casting shops. Another objective of the fabrication is to use the model turbine later for laboratory-scale tests of sediment erosion.

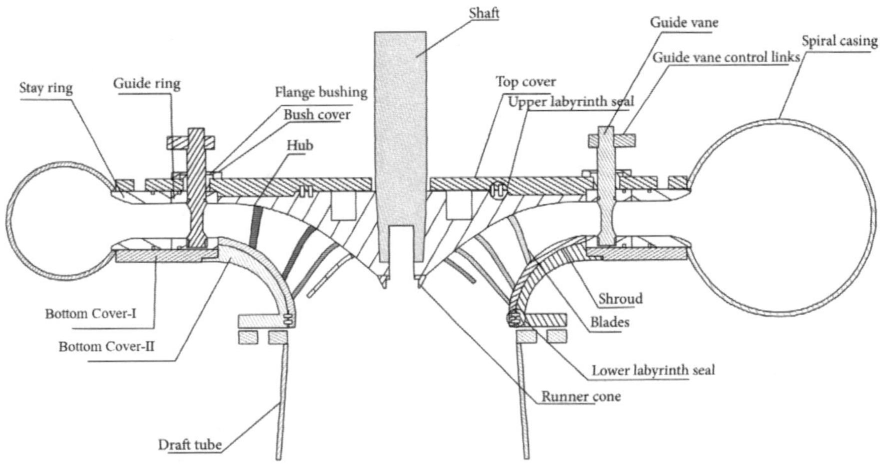

Figure 2. Francis turbine and components [11]

2. Materials and Methods

Pre-existing design of 2 kW model Francis runner available at Turbine Testing Lab was utilized in this study. The existing design was modified by separating the three components hub, shroud and blades of the runner as shown in Figure 3.

Figure 3. 3D printed patterns of shroud, hub (left) and guide vane(right)

The blades were provided with two number of pins along the curve connecting to the shroud part. The shroud was provided with two number of holes for the pins in the blade. This was done to correctly place the blades on the shroud while the runner was fabricated. The separate part files of each component (hub, shroud, and runner) were converted to .stl file for 3D printing. The 3D printers available at the Design Lab of Kathmandu University were utilized for printing the models of the hub shroud, blade, and guide vane.

The 3D-printed blades, hub, and shroud were used as molds for sand casting. The sand casting was done at a local sand-casting shop which specialized in casting brass and bronze sculptures. Photos from local casting shop are shown in Figure 4. Brass was selected for the fabrication of the runner and guide vanes for two specific reasons, (a) Ease of fabrication of brass material with sand casting, and (b) Conducting accelerated erosion testing for studying the erosion in Francis runner.

Figure 4. Sand casting die (left) and pouring molten brass into the die (right) in local sand casting shop.

The metal casting process known as sand casting, or sand mold casting, is characterized by the utilization of sand as the mold material, and with sand referred to as casting sand. Figure 5 displays the process of sand casting. The mold material is made of sand, which is refractory enough even for steel

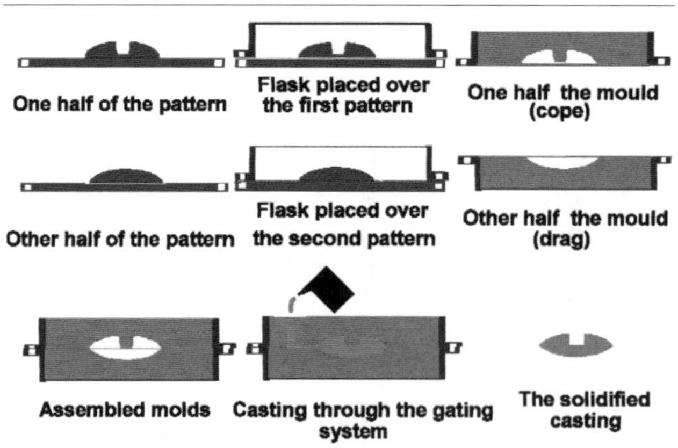

Figure 5. The process of sand casting [13]

foundry applications, and it is accompanied by the use of an appropriate bonding agent [12]. Some local sand foundry also uses molasses as a binding material in the sand mold. During sand casting process first, the pattern or the physical model is pressed in the sand die. Then the pattern is incorporated in the sand and the gating system is made. The pattern is then removed from the sand mold. After this, mold cavity is filled with molten metal and it is allowed for cooling. After the mold has cooled down, Sand mold is removed and the solidified casting is taken out [13].

The guide vane selected for the study was NACA 63(4)-421 profile as shown in Figure 6 and 7. This is an asymmetrical profile with a sharp bent trailing edge. A few millimeters from the trailing edge were trimmed and ground off after casting to remove possible causes of instabilities at the trailing edge.

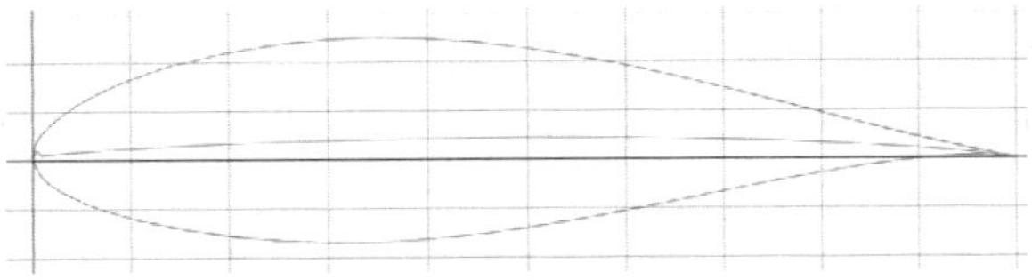

Figure 6. NACA 63(4)-421 profile selected for guide vane

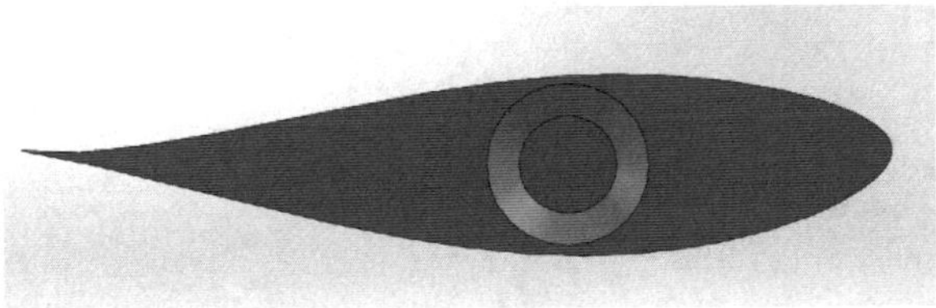

Figure 7. 3D model of guide vane

All the components were cast with brass material except two blades which were cast with copper. This was done to study the variation in the properties of the cast blades. Attempts were also made to fabricate a few guide vanes with copper but with copper, the proper finishing could not be achieved as there was more porosity in the GV cast with copper material. As shown in Figure 8 the GV cast with brass contained less defects and better finishing than that of copper. The cast hub, shroud and guide vanes were machined

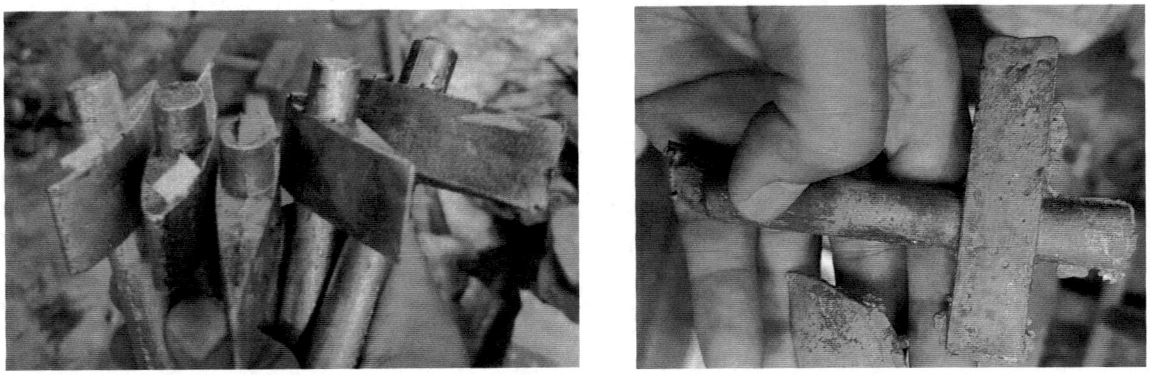

Figure 8. GV cast with brass (left) and GV cast with copper (right)

to remove 2 mm of outer material in order to achieve smooth finishing and to remove the porosities or any other defects encountered in the castings. The runner blades were only polished with a fine polishing wheel. The results after machining and polishing of the runner hub and shroud can be seen in Figure 9. Figure 10 shows the polished runner blades along with machined guide vanes. All the components were

Figure 9. Hub and shroud.

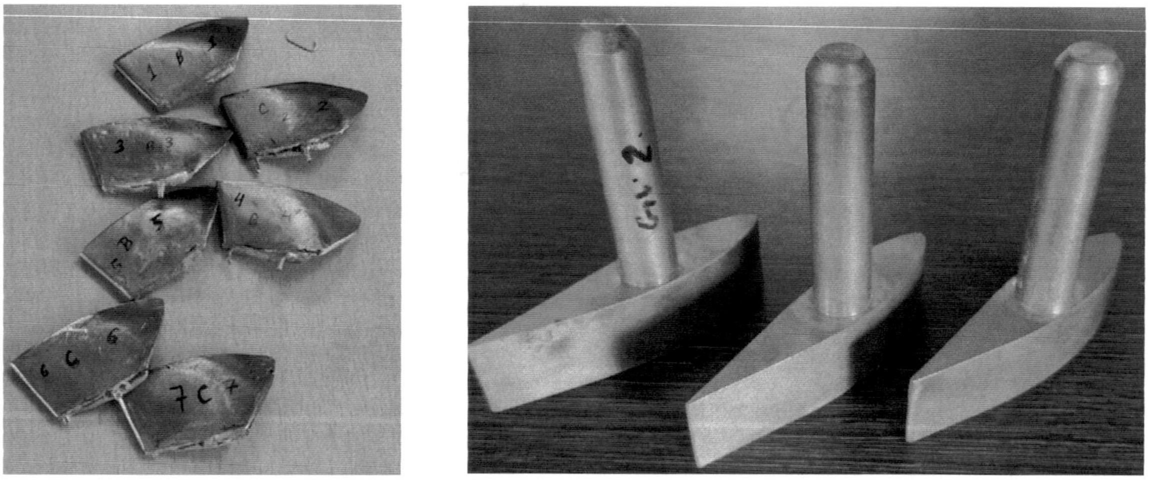

Figure 10. Sand cast and polished runner blades (left) and guide vanes (right).

3D scanned with EinScan HX 3D scanner with an accuracy of 0.1 mm. The 3D scanner was configured in laser scanning mode and calibration of the scanner was done with a calibration board provided by the manufacturer. Markers were attached to the surface of the components to be scanned which served as reference points and helped in transitioning from one point to another during scanning. Upon scanning one side of a component, a new project file was created to scan the other side. The consecutive scans of the different faces were merged into a single file to include all the perspectives. Finally, .stl files were generated for the 3D scans obtained and post-processing was conducted using Geomagic Design X software. Figure 11 shows the photograph of 3D scanning of a guide vane. Finally, after recording the 3D scans of all the components of the runner, the assembly was done. First, the runner blades were assembled with the shroud. The pins in the blades were fitted with the holes in the shroud as shown in Figure 12. Then, the Hub was placed on the top of the assembled blades. Finally, the components were joined together with brazing. A few burrs and splatters occurred during the brazing and were removed with the help of a profile grinder. The final assembly of the runner was obtained as shown in Figure 13. Finally, comparisons were made between the 3D scanned components and the original design. The 3D scanned models of the runner components were converted to the mesh file generated from the point cloud data. This mesh file was then exported in .stl file format. Thus created .stl files were imported into

Figure 11. 3D Scanning

the CloudCompare software along with original 3D designs of each component. The two models, the scanned model and the original model were then aligned with a common origin such that the common features and faces were aligned together. CloudCompare has a feature called cloud registration which registers each point of the meshes in each model. Then another feature of CloudCompare software was used which computes the distance between the registered points of the scanned model and the original model. The original model was selected as the reference against which the distance from the points of the scanned model was calculated. The software then generates a colored contour on the surface of the scanned model that represents the distance for the similar aligned points of the original model. The distance is termed as C2M distance which shows the deviation of the scanned model from the actual design. The comparative analysis is shown in the results section.

Figure 12. Finalized runner.

Performance testing of the sand cast runner was carried out at the erosion testing laboratory of Kathmandu University. The runner's performance was measured in terms of hydraulic efficiency. Pressure data were collected from the inlet of the spiral casing and the draft tube. The measurement of torque was carried out with the help of a torque transducer coupled with the turbine shaft. The difference in pressure readings between the inlet and outlet of the turbine was calculated as net head. The efficiency of the turbine was calculated from equation (1).

$$\eta = \frac{\tau \times \omega}{\rho \times g \times Q \times H} \times 100\%, \tag{1}$$

Where, η is the efficiency of the runner, τ is the torque, ω is the angular velocity, ρ is the density of the fluid, g is the acceleration due to gravity, Q is the flow in m^3/s, and H is the head in meters. The

Figure 13. Erosion testing facility at Turbine Testing Lab, Kathmandu University.

efficiency was measured by keeping the head constant at 4.3 meters and the rpm was varied between 500 to 1200 rpm. The efficiency curves were obtained for three different guide vane angles i.e. 14 degrees (100% opening), 9 degrees (65% opening), and 5 degrees (35% opening). The data from the pressure sensors (inlet and outlet of runner), torque transducer, and flow sensor were logged at a rate of 500 samples per second. The raw data from the sensors were converted to the required unit of pressure by applying the required conversion equation. The data thus obtained were processed via statistical tools as explained. The statistically processed data were averaged to obtain the average value of inlet pressure, outlet pressure, flow, torque, and rpm.

Evaluation of uncertainty in measurements of efficiency was conducted for the experiment. There are two categories of uncertainties in an experiment: Type A and Type B. Type A uncertainty involves assessing measurement uncertainty through a statistical analysis of measured quantities obtained under specific measurement conditions [7]. The reduction of this uncertainty can be achieved by conducting a greater number of measurements. On the other hand, Type B uncertainty stems from sources unrelated to statistical analysis based on a series of observations, and increasing the number of measurements does not mitigate this kind of uncertainty. The efficiency vs rpm curve shown in Figure 14 shows the performance of the runner at the largest GV opening of the experiment whereas developing error bars in the curve shows the uncertainty for each rpm. Maximum total uncertainty in the measurements of efficiency was calculated as 2.675% at 1100 rpm.

3. Results and Discussion

This study has attempted to make qualitative as well as quantitative comparisons between the design and the sand-cast components. CloudCompare software with a feature called C2M signed distance is used for comparing the surface of casted items to the original design. The CloudCompare software is generally used in industries that require visualization and analysis of 3D point cloud processing. It is used in aerospace and automotive industries for quality control of the manufactured products. It is also used in medical imaging and dentistry as well. Similarly, this study has utilized the software for measuring

IOP Conf. Series: Materials Science and Engineering **1314** (2024) 012007 doi:10.1088/1757-899X/1314/1/012007

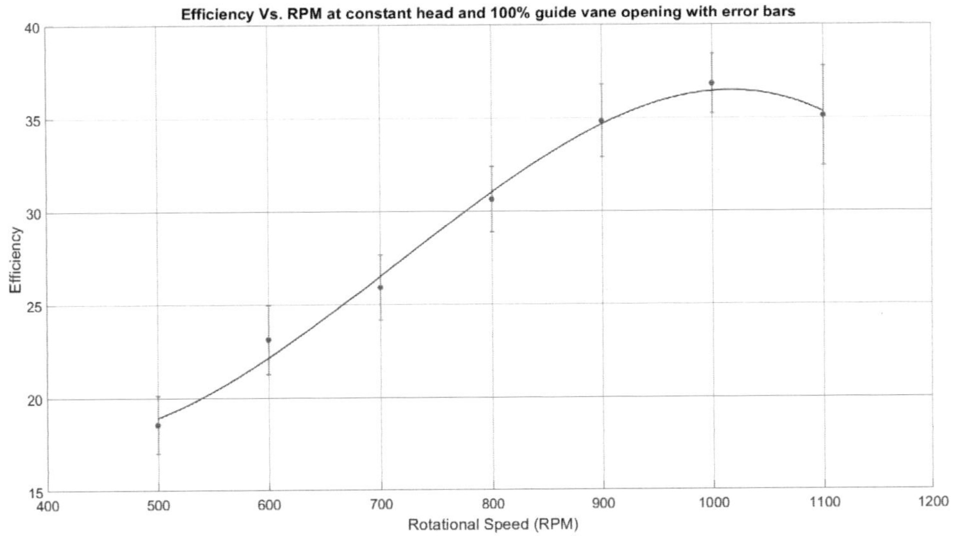

Figure 14. Evaluation of Uncertainty in Efficiency Measurements.

the quality of the sand-cast components of a Francis runner. The deviation in the surface of the blades could result in a loss of efficiency and could even affect the structural integrity of the runner. Figure 15 shows the 3D scan of the cast blade superimposed with the original design. As shown in the figure, the maximum deviation in the computed distances for the runner blades 1 and 2 is observed as 0.78 mm on the positive side and a maximum deviation of -1.34 mm on the negative side. Similarly, the histogram shown in Figure 16 shows the distribution of the number of points with the deviation in distance from the actual design.It demonstrates that the deviation in the profile of the cast blade 1 lies mostly within -1 to 0.5 mm. The contours representing the distance of the mesh points of the cast blade are compared to that of the original design for blades. The areas marked with red and blue contours are the points farthest away from the design points. Thus, it is observed that the method applied in this study along with the sand casting technique for the fabrication of the Francis runner has good accuracy with fewer defects.

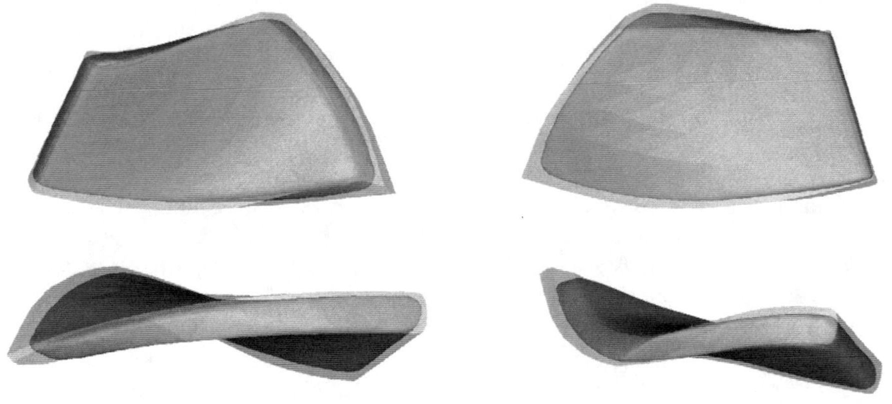

Figure 15. Aligned scanned model (blue color and opaque) and original design (transparent).

The maximum overall deviation in the computed distances among all the components of the runner was found in the case of the runner hub. The comparative analysis for the runner hub is illustrated below in Figure 17. The maximum positive and negative deviations in the runner hub were found to be 2.05

IOP Conf. Series: Materials Science and Engineering **1314** (2024) 012007 doi:10.1088/1757-899X/1314/1/012007

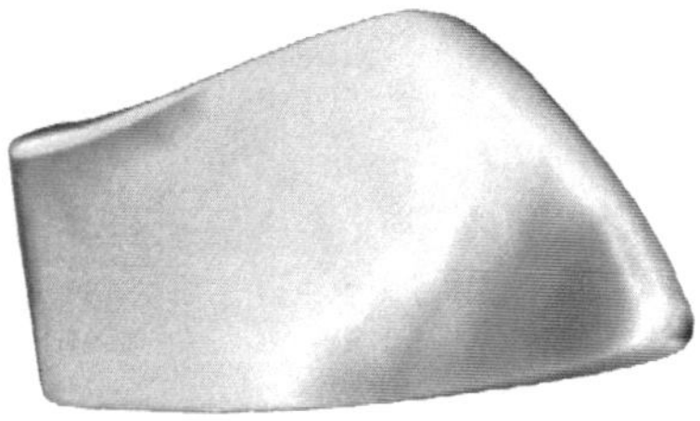

(a) Runner blade no. 1

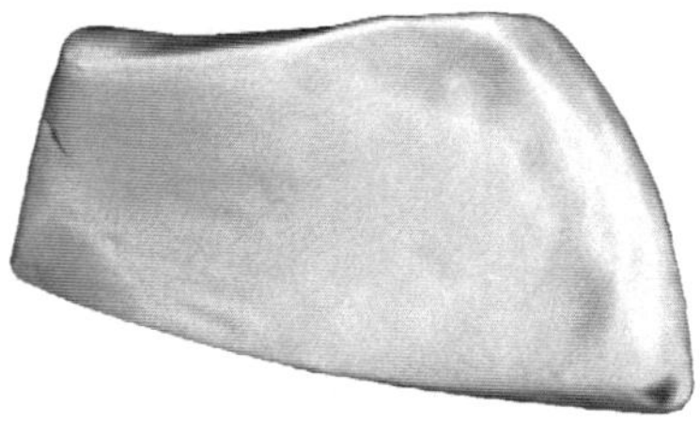

(b) Runner blade no. 2

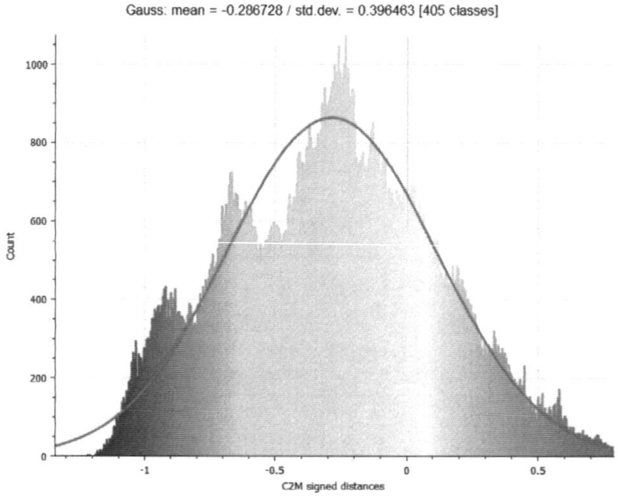

(c) Histogram of computed distances

Figure 16. Contour and histogram showing the computed distances between the points of scanned model and original design of runner blades.

mm and 1.33 mm respectively.

The efficiency curves obtained for three different guide vane angles as shown in Figure 18. As observed from the efficiency curves, the best efficiency of 51.5% is obtained at 35% GV opening. The best efficiency point is different for different openings of guide vanes. Similarly, at 65% and 100% GV opening the maximum efficiency observed was 48 and 36 %. The 2 kW runner was tested at off-design conditions due to the limitations of the experimental setup. The runner was tested at a low head of 4.3 meters. The Q11 and N11 for the three different GV openings were calculated and a hill chart was plotted for these operating conditions. Since the runner was tested at the off-design condition at a low head of 4.3 meters, a partial hill chart was obtained as shown in Figure 19. As the performance tests were carried out at off-design conditions at a relatively low head than that of the designed head, the BEP was obtained at 35% GV opening at 1200 rpm. A similar design and size of the Francis runner was tested in 2012 at Turbine Testing Lab and the maximum efficiency obtained at the design conditions was 52% [6]. Due to the limitations of the test equipment, the sand cast runner produced in this study could not be tested at design conditions.

4. Conclusion

In this study, separate molds for different components of the Francis turbine were produced with 3D printing methods. These printed patterns were used as molds for sand casting the runner components, which were joined together through brazing. This method proved to be efficient in terms of both speed and cost of manufacturing as compared to traditional casting methods. Furthermore, this method of

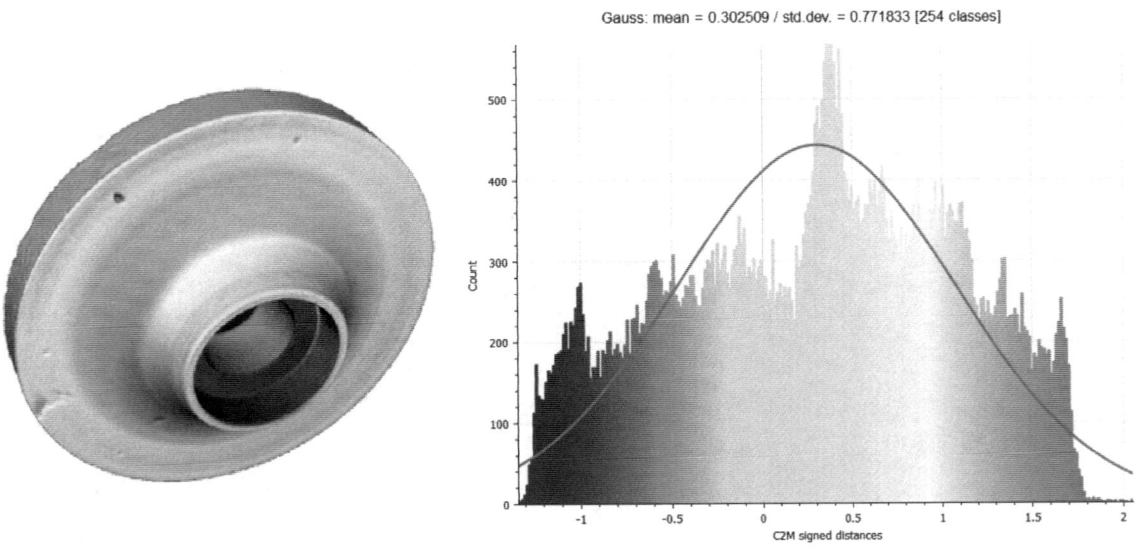

Figure 17. Contour and histogram showing the computed distances between the points of scanned model and original design of runner hub.

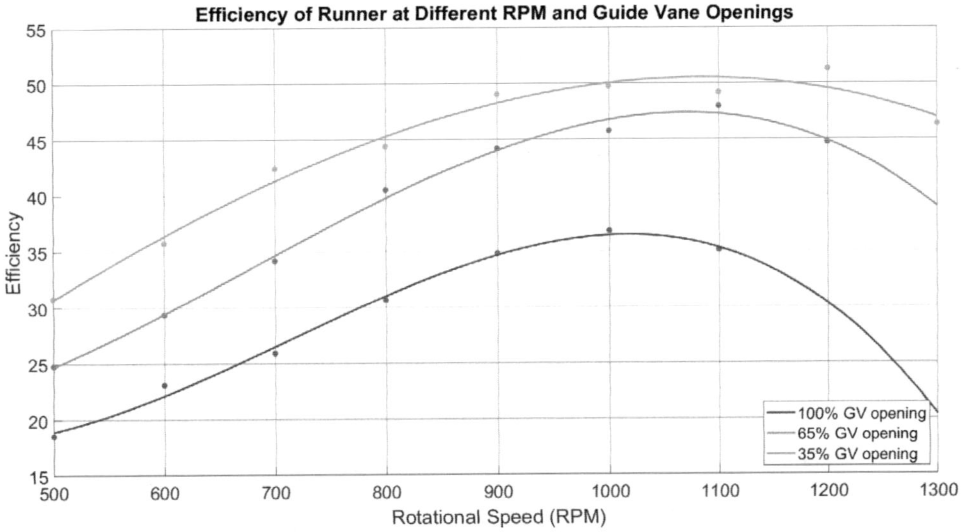

Figure 18. Rpm vs. Efficiency at different G.V. opening (constant head efficiency curves)

manufacturing produced fewer defects in runner components compared with single-piece cast runners with investment casting. 3D scans of the runner components were conducted using the EinScan HX 3D scanner and the accuracy of the fabricated model was assessed. The 3D scans of the fabricated components were compared to the original design and the maximum deviation was found to be 2.05 mm in the case of the runner hub. The deviation in the profiles of runner blades was a maximum of 1.5 mm. The best efficiency for the sand cast runner was obtained as 51.5% at 35% GV opening. Similarly, at 65% and 100% GV opening the maximum efficiency observed are 48 and 36 %. The 2 kW runner was tested at off-design conditions due to the limitations of the experimental setup. The runner was tested at a low head of 4.3 meters.

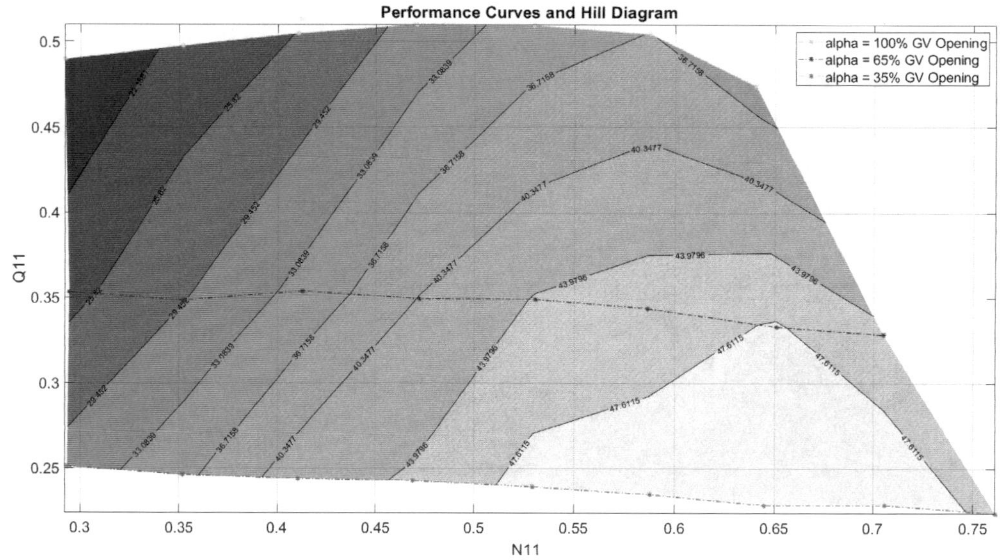

Figure 19. Hill Diagram for 100%, 65%, and 35% opening of guide vane opening

To conclude, the fabrication method for the Francis turbine which utilized 3D printing and component-wise sand casting proved to be effective in terms of cost and time required for fabrication. The method was also successful in reducing defects in the cast. This method could serve as an alternative for smaller-scale turbine production. Moreover, Further research will be necessary for full-scale commercial applications of this method using high-quality alloy steel.

Acknowledgments
The study was performed jointly at the Turbine Testing Lab and Design Lab of Kathmandu University and State Key Laboratory of Water Resources and Management, Wuhan University. The project was funded by the Visiting Scholar's Fund provided by WRHES, China (Project No. 2021SDG01), Hydro Himalaya Project, Nepal, and EnergizeNepal (ENEP-RENP-II-23-03) program. Authors express gratitude towards Aatmaram Kayastha, Bhuwan Bhattarai, Rakish Shrestha, and Prithvi Gurung for their support during the project.

References
[1] Thapa B 2004 Sand Erosioni in Hydraulic Machinery Ph.d. thesis Norwegian University of Science and Technology Trondheim
[2] NHE 1999 Core Business Area and Capability URL `https://nhe.com.np/capability`
[3] Thapa R, Thapa B, Thapa B SLee Y 2019 Learning the Lesson from Maintenance of Turbine due to Sand Erosion and Utilize that Knowledge for Manufacturing New Turbines in Nepal
[4] Feasibility Study for Francis Turbine Manufacturing and Testing Facility in Nepal
[5] Koirala R, Chitrakar S, Panthee A, Neopane H PThapa B 2015 Implementation of Computer Aided Engineering for Francis Turbine Development in Nepal *International Journal of Manufacturing Engineering* **2015**
[6] Thapa B S 2012 Introduction to TTL, its Activities and Future Cooperation
[7] Dahal D R 2019 Design of Francis Turbine and Propose its Manufacturing Procedure in Context of Himalayan Region
[8] Mathema S 2015 Private Sector in the Development of Nepal's Micro Hydro
[9] Kafle A 2022 Development of Procedures for Manufacturing of Francis Turbine Runner in Nepal
[10] Chitrakar S 2022 Introduction to Francis Turbine

[11] Koirala R, Neopane H P, Zhu BThapa B 2019 Effect of Sediment Erosion on Flow Around Guide Vanes of Francis Turbine *Renewable Energy* **136** 1022–1027
[12] Beeley P 2001 Foundry Technology (Elsevier)
[13] Chelladurai C, Mohan N, Hariharashayee D, Manikandan SSivaperumal P 2021 Analyzing the Casting Defects in Small Scale Casting Industry *Materials Today: Proceedings* **37** 386–394

Design of Force Sensitive Resistor (FSR) embedded insole for phase detection during human gait and its classification

U Pant*[1], S Baral[1], A Gupta[1], P L Shrestha[1]

[1]Design Lab, Department of Mechanical Engineering, Kathmandu University, Dhulikhel, Nepal

*Corresponding address: upamapanta@gmail.com

Abstract. Gait analysis is the study of human motion during walking or running. It is often used to evaluate and treat individuals with movement disorders or to assist in the design of prosthetics and other assistive devices. The phase detection aspect of gait allows for the quantification of gait characteristics, assessment of effectiveness of interventions and monitoring the changes in gait over time. The objective of the paper is the design and development of a Force Sensing Resistor (FSR) embedded insole that can be placed in shoes for detection of phases during the human gait and its classification. The classification using FSR is achieved by placing sensors at strategic locations on the foot, to detect the forces applied during walking. The data collected by these sensors are used to identify specific phases of the gait cycle—particularly heel strike, midstance, toe-off and swing, by analysing the patterns of data during movement. The layout and number of FSRs is determined through literature review and is followed by the insole fabrication. The wireless transmission of gait data through the FSRs is actuated for ten healthy adults. Machine Learning (ML) program is consequently developed based on all available classification algorithms of supervised learning namely—Logistic Regression, Naïve Bayes, Decision Tree, Random Forest and Support Vector Machine (SVM). The algorithm is trained and tested on separate FSR data sets from which the predictive decision in each gait cycle is attained. In the phase classification aspect, SVM, Naïve Bayes, and Random Forest achieved the highest accuracy of 91.54%, followed by Decision Tree at 90.77% and Logistic Regression at 89.23%. In conclusion, the optimal number of FSRs and their location in smart insole for gait phase detection is resolved. Gait phase classification is also attained. FSR embedded insole thus, provides a promising solution in accurately identifying and categorizing gait phases.

Keywords: FSR, insole, gait analysis, phase classification, machine learning

1. Introduction

Gait refers to the manner or style in which a person's mobility is defined. The information can range from intent, gender to medical conditions and abnormalities. Gait analysis is defined as the inspection and assessment of human locomotion [1]. Gait analysis can be used for identifying congenital clinical abnormalities such as Parkinson's disease, Cerebral Palsy, Huntington's disease that develop within. It causes inadvertent movements like shaking, stiffness and disorientation [2]. Similarly, it can also be devised for gender determination. Moreover, gait analysis can be used to identify acquired conditions such as musculoskeletal conditions. It includes conditions such as osteoarthritis where a person experiences stiffness during the motion of the joints [3]. Gait analysis can be used in emotion recognitions too where reduced walking pace, arm swing, and vertical head movements are characteristics of gait patterns linked to melancholy and depression [4].

The isolation of shortest, unique and repeatable gait task is the preliminary step in the quantification of the walk [5]. The gait cycle consists of two main phases which is the stance phase and the swing phase. Each of these phases consists of further stages that are illustrated (Figure 1) [6].

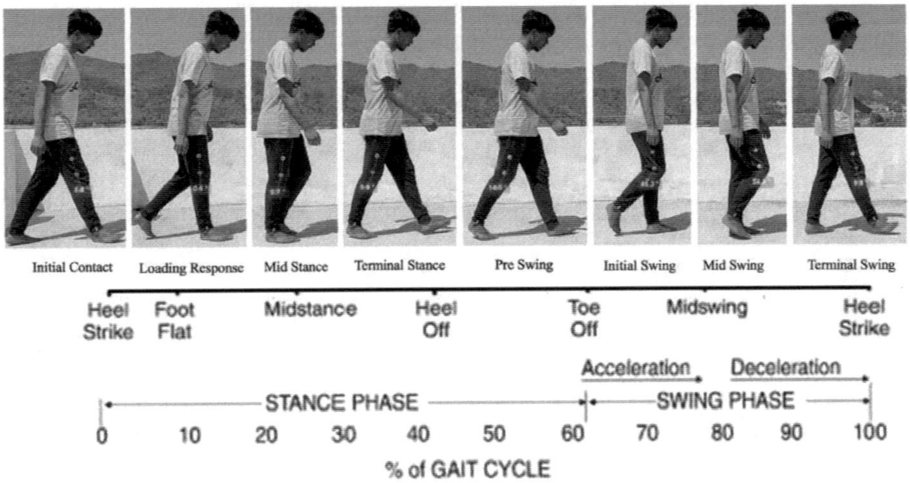

Figure 1. Phases of gait cycle

Kinetic analysis based on GRF is frequently used in gait studies, where GRF corresponds to the person's weight at rest and increases with movement [7]. However, GRF is not essential for gait phase detection. Traditional gait studies in motion analysis laboratories often utilize force plates and infrared cameras, but these setups are costly and complex [8]. Therefore, more portable and cost-effective methods, such as wearable sensors, pressure-sensitive insoles, and marker-based motion capture systems, are preferred [9]. "INSOLE-BASED GAIT ANALYSIS" [10] has focused on affordable gait analysis systems, with insole-based solutions emerging as a viable option. These systems, which incorporate FSRs, offer respectable accuracy, reduced sensor requirements, and easier accessibility. Despite this, the optimal number and placement of sensors for effective gait phase classification remain underexplored.

Often systems that operate in providing gait analysis orbit around exorbitant prices, space requirements and the demand of highly specialized personnel for their operation [11]. To address this limitation, wearable technology can prove to be a much accessible gait analysis technique. The objective of this study is to design FSR embedded insole and train ML classifiers to understand its application in gait phase detection and finally discuss the results obtained. As the data obtained during gait analysis is too big, it is effective to find solutions in Machine Learning to detect gait phase. In a research by Yang et. Al. three methods 1) Logistic Regression (LR), 2) Random Forest Classifier (RF), and 3) Artificial Neural Network (NN) are used to build the gait phase detection models for and IMU based portable gait phase detection unit [12]. Similarly, in a research by Rezaei et. al. three commonly used machine learning algorithms (random forest, support vector machines, and neural network) were investigated to differentiate between four phases (initial contact, mid-stance, pre-swing, swing) of the walking data. The random forest classifier performed best followed by neural network and support vector machines. Using to-fold cross-validation, an average accuracy of 95.49% ± 2.32% was achieved [13].

Moreover, this research aims to identify stress points on the insole for the optimum number of FSRs required along with their ideal placement. The research also intends to acquire gait data from complete gait cycle of a set of participants who have no prior gait condition. Finally, this research intends to produce a graphical representation of individual FSR readings and classification of four gait phases using ML.

2. Materials and methods

2.1. Gait phase division

The walking cycle is described in the biomechanics of the march as the series of actions that take place in between two repetitions of the same heel making contact with the ground [14]. The graph (Figure 2) [15] displays the changes in plantar pressure throughout the gait cycle whereas the graph (Figure 3) [16] displays the changes in the total plantar pressure during the stance phase.

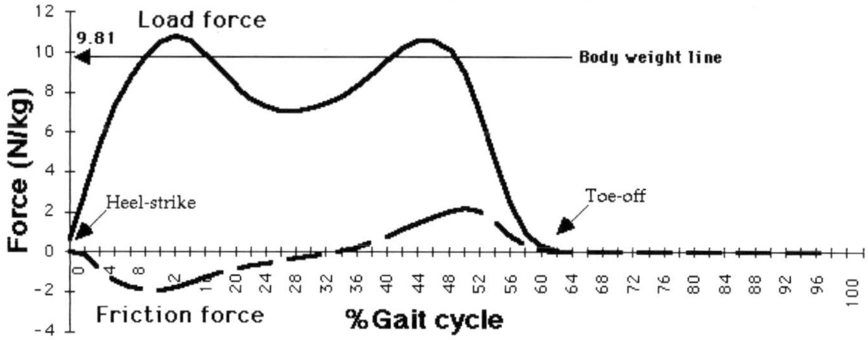

Figure 2. GRF during a complete gait cycle

It is inferred that maximum pressure vacillates between heel strike and toe off which are enough to determine gait phases. Change in continuous values aid in further determination of stance phase too. Three sub phases of stance phase, i.e. heel strike, mid-stance and toe-off, and swing phase are the designated gait phase assigned for detecting gait.

Secondly, as the body moves from heel strike to toe-off, there is a gradual shift in the body's center of mass from the posterior to the anterior portion of the foot. This shift in weight distribution causes a greater force to be applied to the toes and forefoot, resulting in a higher GRF during the toe-off phase.

Lastly, during the toe-off phase, the muscles in the foot and leg undergo a significant concentric contraction, generating a large propulsive force that helps to lift the body off the ground. This propulsive force also contributes to the higher GRF observed during toe-off [17, 18].

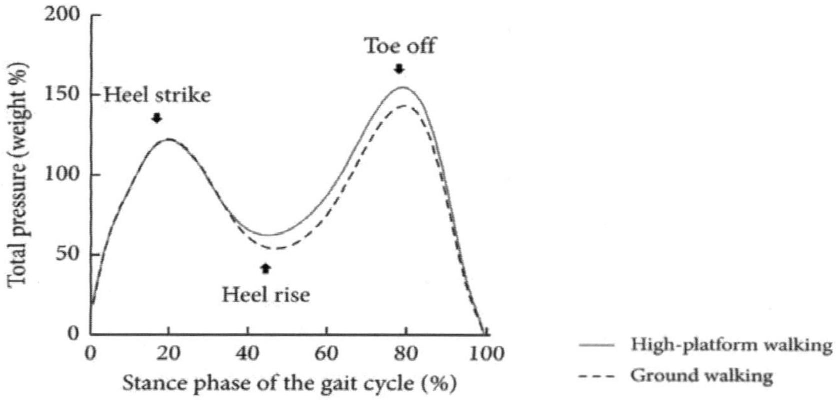

Figure 3. Change in average total plantar pressure during stance phase

IOP Conf. Series: Materials Science and Engineering **1314** (2024) 012008 doi:10.1088/1757-899X/1314/1/012008

2.2. FSR embedded insole

FSRs are sensors that can recognize weight, squeezing, and physical pressure. They are inexpensive and easy to use. The sensitive portion is the circular piece with a 1/2" diameter [19]. The FSRs, specially the Interlink 402 (Interlink Electronics) is about 0.6mm thick and hence is easy to use in insole as it causes minimal discomfort [20]. To fabricate the insole, EVA was used as it showed better performance in comparison to PDMS, TPU and FRP in terms of accessibility, cost, stiffness, method of processing, durability and comfort of wearing [21, 22, 23, 24] [25] [26, 27, 28] [29]. To determine number and location of FSR embedded in the insole, literature review on prevalent researches was done (Table 1).

Table 1. Review of the number of FSRs used and their placement

Reference number	Number of FSR	Location description	Picture
[30]	4	One in proximal phalange, one in metatarsal bone and one in calcaneus.	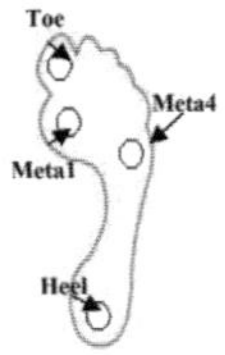
[31]	5	One in middle phalange, one in proximal phalange, two in lateral metatarsal bone and one in calcaneus.	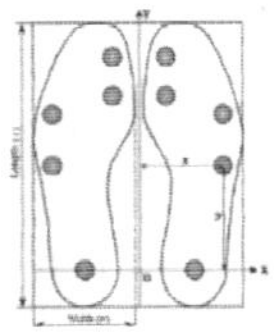
[32]	4	One in distal phalange, one in proximal phalange, one in lateral metatarsal and one in calcaneus.	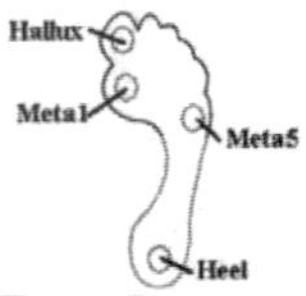
[33]	4	One in distal phalange, one in proximal phalange and two in calcaneus.	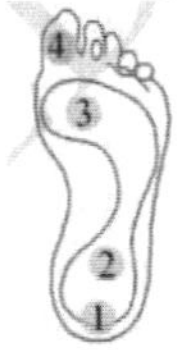
[34]	3	Two metatarsal bones and one calcaneus.	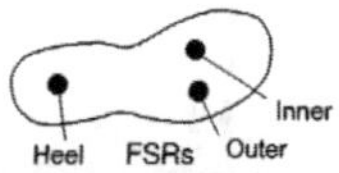
[35]	16	Distributed over the entire foot.	
[36]	2	One in proximal metatarsal and one in calcaneus.	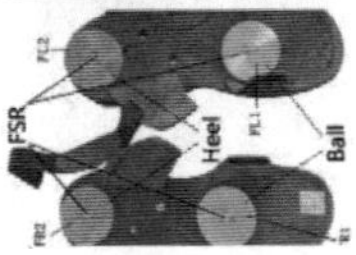
[37]	4	One in distal phalange, proximal phalange, one in metatarsal bone and one in calcaneus.	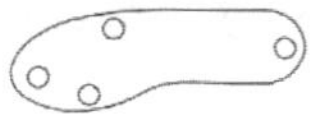

[38]	9	Distributed over the distal, proximal, metatarsal and calcaneus region.	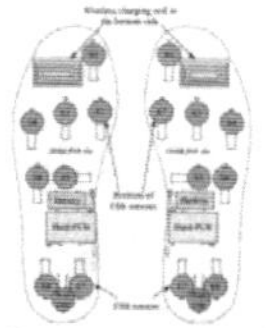
[39]	8	Distributed over the entire foot.	
[40]	2	One in proximal phalangeal and one in calcaneus.	
[41]	4	One in distal phalange, two in proximal phalange and one in calcaneus.	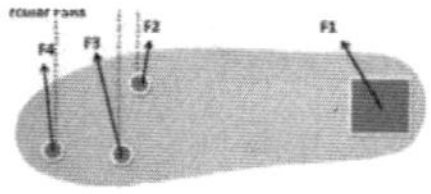

The architecture of the foot, illnesses, and different types of footprints (pronator, supinator, neutral) are all taken into consideration when deciding where to place the sensors. The heel (regions 1-3), the arch (regions 4-5), the metatarsus (regions 6-10), and the fingers (areas 11–15), as indicated in Figure 8 (left), comprise the 15 areas residing in the sole.

From Table 1, locations 1, 5 and 10 (one in each region) as taken from Figure 8 (right) have been deemed appropriate for the incorporation of three FSRs (one in each region) for the integration in the sensing region of the insole. The intent is to achieve gait phase classification with least number of FSRs in the insole.

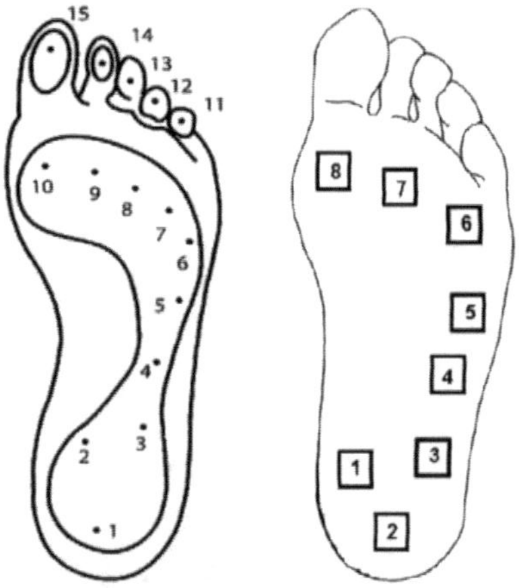

Figure 4. Highest points of foot pressure (left), Optimal choices for FSR location (right)

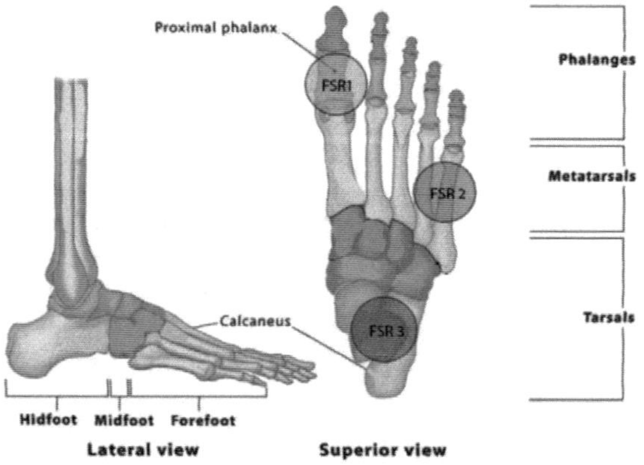

Figure 5. Location of FSR corresponding to the anatomy of the human foot

The FSRs are strategically placed in the proximal phalangeal, lateral metatarsal, and calcaneus, as illustrated (Figure 4) [42]. The FSR located in the heel region is intended to detect the onset of the heel strike, while the one placed in the arch region is expected to identify the occurrence of supination. Finally, the FSR embedded in the forefoot region is designed to detect the toe-off event, which marks the end of the stance phase and the initiation of the swing phase. In the swing phase, all FSR readings become null.

IOP Conf. Series: Materials Science and Engineering **1314** (2024) 012008 doi:10.1088/1757-899X/1314/1/012008

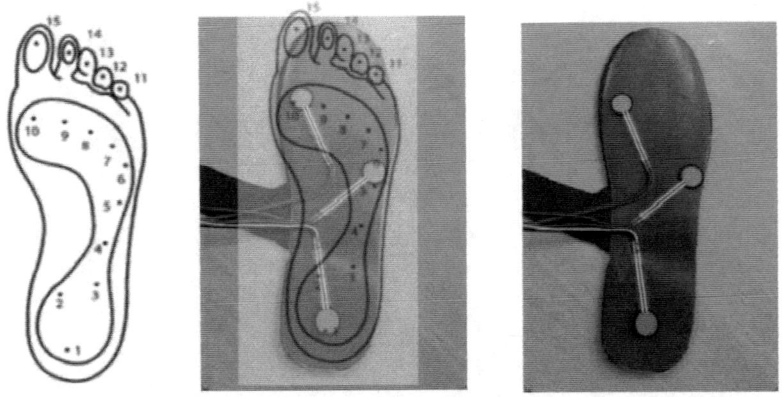

Figure 6. FSR placement in accordance with the areas of the foot corresponding to highest pressure as displayed in Figure 5

Considering the literature review on gait analysis using FSR, the number and location of FSR was finalized as shown in figure 6. The corresponding FSR locations lie in the displayed circular range (Figure 4). A circular area is identified to accommodate foot size variability.

2.3. Data acquisition and transmission
The resistors in use were of 5 KΩ each. The resistors in the circuit help to limit the current flow and prevent damage to the circuit by limiting the maximum voltage that can be applied to the FSR. Additionally, the use of resistors allows for the signal to be amplified and filtered, resulting in a more accurate measurement of the applied force [43]. Testing was implicated by applying force to the sensors and observing the output of the Arduino in the serial monitor. The gait data was taken from 10 individuals, 4 females and 6 males with no prior gait conditions. The anthropometric data consists of participants' demographic with age 22-24 years, weight: 44-65 kg, height: 4'11"-5'10".

The MCU in use is the DOIT ESP-32 Devkit V1. The unit consists of 12-bit (0-4095 digital signals) SAR ADC with 18 channels where the FSR signals are drawn out towards. The continuous analog values are displayed in the serial monitor as digital values. The Wi-Fi module is 802.11 b/g/n [44]. The pinout of the MCU along with the circuit is given (Figure 7). The data from the insole was transmitted using the Thingspeak® platform developed by MathWorks, where a channel is setup to receive data from the ESP-32. Figure 8 shows the circuit connection we used for FSR, power source and ESP-32. Similarly, Figure 8 is the final fabricated set up FSR embedded insole with ESP-32 in the right shoe.

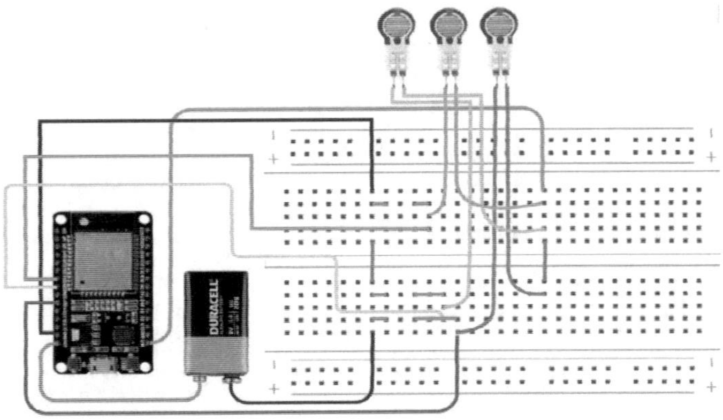

Figure 7. Connection diagram of ESP-32, FSRs on a breadboard

Figure 8. Assemblage of the circuit and the insole in a shoe

2.4. Formulation of gait phase detection algorithm

The heel strike and toe-off events are the most crucial characteristic points for wearable-based gait analysis. These occurrences, which correspond to the stance phase and swing phase of the human gait represent salient and, hence, clearly recognizable indications in wearable sensor data [45]. The idea is to concentrate on the identification of heel strike, mid-stance, toe-off as events of the stance phase as well as the swing phase because the FSR-based gait segmentation is the primary scope of this work. The segmentation algorithm receives its input signal from the summed-up FSR time series [46]. The performance of the developed system has been estimated for five SVM classifiers namely—Logistic Regression, Naïve Bayes, Random Forest, Decision Tree and Support Vector machines.

The variability in the numerous parameters along with the volume of data involved in conducting gait analysis make the task of interpretation taxing [47, 48]. The application of machine learning in integration with the aspect of biomechanics is thus promising. The ML model is trained on a specific data input set and tested on a separate set so that the accuracy of the model is reliable. Algorithms are trained using statistical techniques to produce classifications or predictions and to extract the insights of the data. Decision-making is subsequently driven by these findings [49]. The classification of machine learning is done in the figure below:

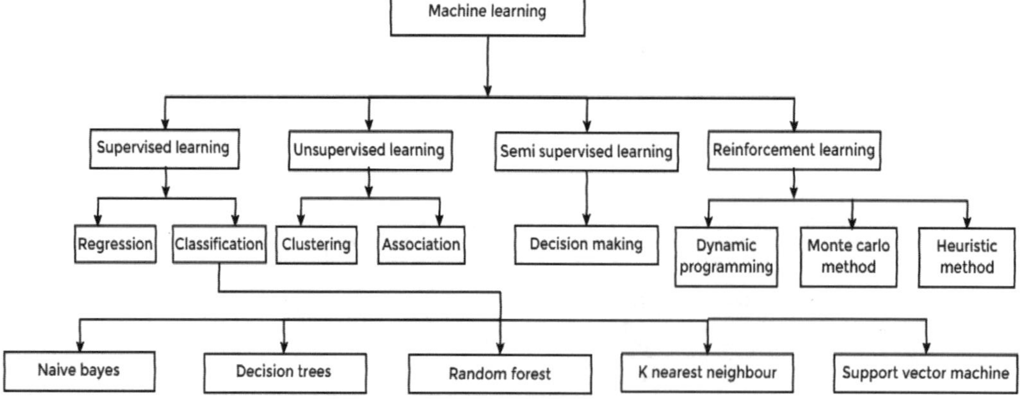

Figure 9. Types of machine learning algorithms

Supervised learning refers to the type where the output labels are already defined and both the input and output of the data are specified. Unsupervised learning involves data that train on unlabeled data. The ML model thus search for any meaning connection present in the same. Semi supervised learning involves a mix of the previous two techniques with a segregation between labeled and unlabeled data. Reinforcement learning is trained through a series of rewards and punishment in corresponding to desirable and undesirable behavior respectively. the environment of data is interpreted by a series of trial and error [50, 51].

Since the gait cycle is already attained, the four phases heel strike, midstance, toe-off and swing can be visualized. The label is categorically given as input by comprehending the combination values of FSR 1, 2 and 3. Since the output is already defined, supervised learning is relevant to make predictions that map the input with the output [52]. The intent is data classification. Additionally, supervised learning involves two different approaches [53]:

i. Regression: Regression algorithm are utilized to predict continuous values based on input variable

ii. Classification: Classification algorithms are utilized to identify discrete output variables that can be segregated as labels or categories. Here, the cardinal difference is that regression predicts a quantity and classification predicts a label [54, 55].

Multi class classification is used to predict gait phase with four outcomes. All available classification algorithms are applied to train the data set for the gait phase classification. The linear model comprises of logistic regression and Support Vector Machines (SVM) whereas the non-linear model consists of Naïve Bayes, Decision tree and Random Forest [56]. They are compared based on the accuracy of their prediction on a data set not fed to the ML model for training purpose. They are as follows [57, 58, 59]:

i. Logistic regression: Logistic regression uses a sigmoid function to return the probability of the layer. While it is generally used in binary classification scenarios, even in multi class, the outcome probability is compared to the predefined threshold. Then the object is assigned to the label.

ii. Naïve Bayes: Based on Bayes' theorem, Naïve Bayes determines the probability of each outcome. The naïve assumption is that each feature is independent to each other.

iii. Decision Tree: The working mechanism of decision tree is comparative to that of a flow chart, utilizing a hierarchical approach to determine which category the input falls into. The categorical decision is made at the leaves, while the data is partitioned through branches.

iv. Random Forest: intuitively, random forest is an assemblage of decision trees. Multiple predictors (the decision trees) a run simultaneously to access an aggregated result.

v. Support Vector Machine: SVM creates a hyperplane in n-dimension such that n is the number of clusters where the plane segregates the data. This plane maximizes the distance between data points from different classes.

The steps involved in formulating gait phase detection algorithm are

1. Data preprocessing: This includes steps like cleaning the data, handling missing values, encoding categorical variables, and scaling/normalizing the data. The panda library which is used primarily for data analysis purposes is also suitable for raw data refinement and the rectification of present anomalies. Any excess columns and rows with strings were removed beforehand as the model demands numeric values.
2. Choice of algorithm in use: The understanding of the algorithms that are suitable for the problem at hand is crucial. The classifiers in use have been mentioned above.
3. Data segregation, training and evaluation: The dataset is split into two subsets namely—training and testing. This is done to provide an unbiased evaluation of a final model fit on the training data set [60].

3. Result and discussion

3.1. Gait cycle graph

Figure 8 displays the gait of subject with Sample ID 1 as extended over five gait cycles. The graph displays the sensor values where it can be seen the FSR 1 located in the proximal phalange has a rather

distributed weight load without any cut off at the peak. FSR 2 located in the lateral metatarsal bone shows discontinuities within the gait cycle and consists of inconsistency in data capture. This is primarily due to the variations in insole fitting within weight class of the participants. FSR 3 located in the calcaneus region has the highest weight load due to which the digital readings have peaked to 4095. This way for the participant weight range of 44-65 kg, three FSR proved to be sufficient.
The FSR trend for three sensors whose placements is actuated in the aforementioned regions of Figure 11 have been compared with the result provided by Kardos S. in "Sensor Array for Evaluation of Gait Cycle" [81].
The anticipated and actual relative readings of FSR have been tabulated in Table 2.

Table 2 FSR readings with respect to gait phase in comparison with expected and actual values

Phase	Expected value			Actual value		
	FSR 1	FSR 2	FSR 3	FSR 1	FSR 2	FSR 3
Heel strike	null	null	high	null	null	high
Midstance	medium	high	medium	medium	high	medium
Toe-off	high	low	null	high	medium	low
Swing	null	null	null	null	null	low

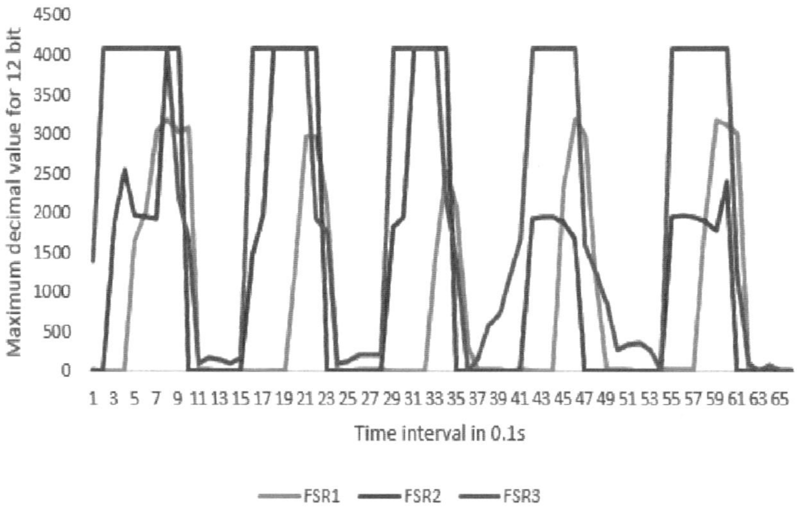

Figure 10. Individual FSR readings of subject with Sample ID 1 comprising of five gait cycles

The gait graph extended over five cycles have been integrated and overlapped over a single cycle with respect to the FSRs in use. The trigger points of each FSR have been noted for each cycle such that they initiate at the same point in the graph. The variations cause by anthropometric as well as spatio-temporal data can be visualized.

From the overlapped plots for Sample ID 1, an average is drawn out to indicated the trend of FSR values over a gait cycle as displayed (Figure 9). The graph allows for a detailed investigation of gait patterns beyond conventional approaches. By examining the shape, magnitude, and timing of FSR readings, hidden features related to gait symmetry, load distribution, gait phases, propulsive forces, and gait abnormalities can be explored.

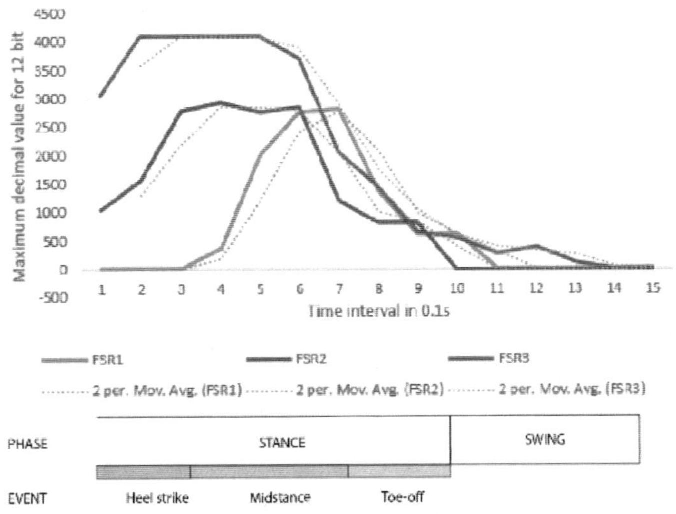

Figure 11. Average readings of FSR 1 (depicted by the green plot), FSR 2 (depicted by the red plot) and FSR 3 (depicted by the blue plot) for subject with Sample ID 1 over five gait cycles

Likewise, such average graphs were drawn out for all Sample IDs which resulted in similar trend. In the graph above, a gait cycle can be inferred from heel strike where FSR 3 makes the first contact followed by FSR 2 and after a delay of 3ms, time required to shift COG from heel to the toe, FSR 1 activates. The plateau of sensor data shows mid-stance. The toe sensor maintains its peak right till the heel sensor value declines showing toe-off. Almost simultaneously the sensors relay null value signifying swing phase. And this cycle repeats producing a similar result for all of the Sample IDs.

3.2. ML classification algorithm

Supervised ML classification algorithms have varying degrees of performance accuracy owing to a plethora of complexities associated with algorithm design, the data's nature, and the underlying mechanisms of the algorithm. The behaviour of each algorithm is influenced by its distinctive features and assumptions, which enable it to capture complex patterns within the data. Additionally, the algorithm's underlying mathematical model, its optimization process, and its tolerance to error and noise can influence its accuracy. Another factor that can impact accuracy is the complexity [61].

Table 3. ML classification accuracy

Machine learning classification algorithm	Total training data	Total test data	Verification sample	Accuracy
Logistic Regression				0.8923
Gaussian Naïve Bayes	517	130	92	0.9154
Decision Tree				0.9077
Random Forest				0.9154
Support Vector Machines				0.9154

Table 3 shows the accuracy of five different machine learning classification algorithms on a gait cycle dataset of ten participants. SVM, Naïve Bayes, and Random Forest achieved the highest accuracy of 91.54%, followed by Decision Tree at 90.77% and Logistic Regression at 89.23%. The results suggest that SVM, Naïve Bayes and Random Forest to be equivalently accurate for classifying instances. The accuracy of classifiers may differ with changing number of training datasets. More number of datasets yield in better accuracies in gait phase detection.

3.3. Classification accuracy and confusion matrix

Classification accuracy is the ratio of correct predictions to total predictions made. On the other hand, the confusion matrix generates a tabulated information of the performance of the ML model [62]. It displays the correct and incorrect predictions made along with the display of count broken by each class.

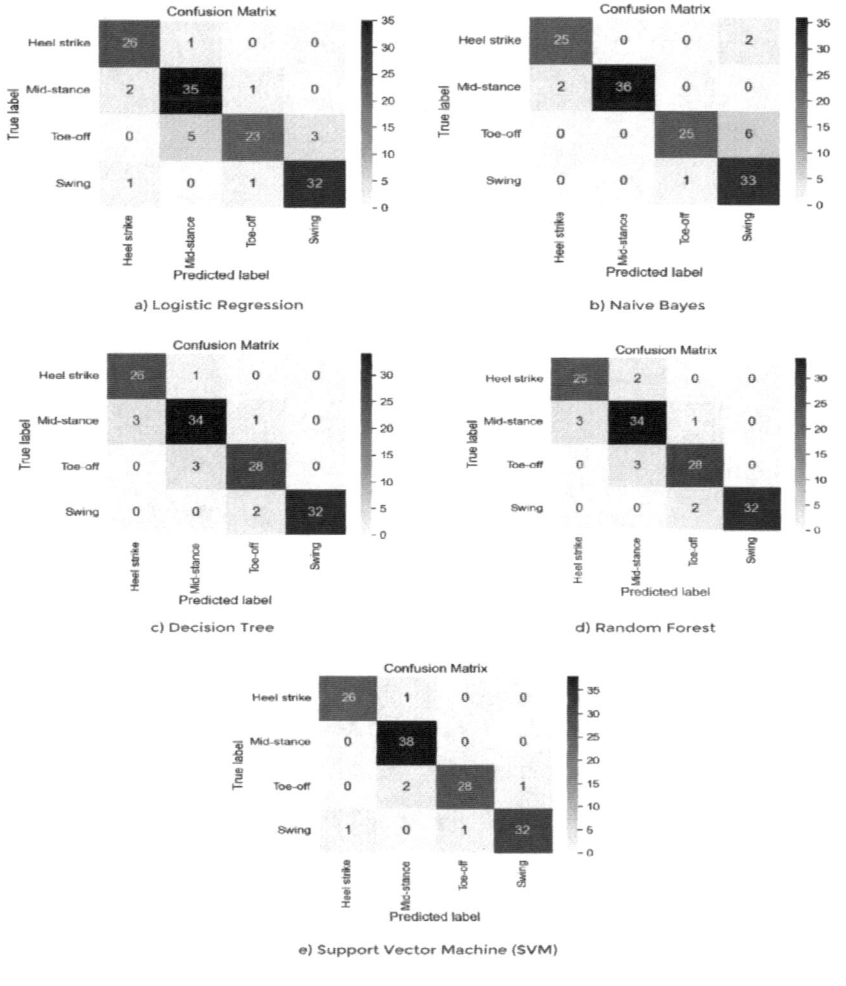

Figure 12. Confusion matrix of the classification algorithms in use

These metrics provide insights into the model's performance, highlighting its strengths and weaknesses in distinguishing between various gait patterns. By analysing the confusion matrix, researchers and practitioners can identify specific areas where the model may be misclassifying gait patterns and adjust the model accordingly. It aids in refining and optimizing the ML algorithm for gait analysis, ultimately leading to more accurate and reliable predictions.

The limitations of the accuracy score of each model is broken down by the confusion matrix such that the insight of error as well as its type is given. The graph above enables the identification of four metrics: true positives, true negatives, false positives, and false negatives.

4. Conclusion

Understanding gait analysis and by analysing foot pressure distribution, FSR embedded insole was designed. In this study, foot pressure distribution and areas of high-pressure concentrations was determined to quantify and locate FSRs in the insole. The FSR readings were collected through the ESP 32 with a program sketch on Arduino IDE itself. The distinct differentiation between the applied pressure and proportional increase in the numeric value of the FSR has been detected. Due to this variation, the evaluation of ground reaction force for the phase detection had deemed to be redundant. The FSR readings are displayed as the 12-bit decimal value of the ADC converter in ESP 32.

Sample data were obtained from a control set of ten healthy adults. The individual graph of these sensors was plotted. For over five cycles for each participant, sensor readings for each cycle were overlapped and exported as average to evaluate the presence of trend. Gait phase was detected and classified using five classification algorithms as a part of supervised learning. The outcomes were incorporated in alignment with the extracted graph. The accuracy was obtained by testing a separate dataset not incorporated in the training data. For a limited data set of 647 values, the accuracy of the model has ranged to a peak value of 91.54 %.

5. Recommendations

The research was focused in fabrication of an FSR embedded insole and gait phase detection for which supervised machine learning models were used. Moving forward, there are a few limitations our research faced that could be continued in the future. Firstly, three FSRs were used for the purpose of gait phase detection. However, as the weight class elevated during the data acquisition process, the FSRs were cut off during the stance phase. This was caused due to improper load variation. The FSR count can be increased accordingly from this base value. Moreover, a data visualization platform can be created to tackle the lag issue instead of using a third-party application. Similarly, the accuracy of the ML program depends on the amount of data used for training. Therefore, apt data collection is to be prioritized. Also, the model's performance could be evaluated under a wider range of gait conditions to ensure its effectiveness in a variety of situations. This may involve testing the model on different walking speeds or inclines. Additionally, since three algorithms have shown the same accuracy, the data set used for training the model should be extensively substantial to cater to miniscule variations throughout the gait cycle. Finally, for studies pertaining to specific gait abnormalities and disorders, FSR location must be altered accordingly.

References

[1] M. J. Schepers and R. G. Breedveld, "Gait analysis: IEEE Standard 1057-2007," *IEEE Standards Dictionary Online,* 2021.

[2] "Parkinson's Disease: Causes, Symptoms, and Treatments," National Institute on Aging , 14 April 2022. [Online]. Available: https://www.nia.nih.gov/health/parkinsons-disease#:~:text=Parkinson%27s%20disease%20is%20a%20brain,have%20difficulty%20walking%20and%20talking.. [Accessed 5 July 2022].

[3] M. clinic, "mayoclinic.org," [Online]. Available: https://www.mayoclinic.org/diseases-conditions/osteoarthritis/symptoms-causes/syc-

20351925#:~:text=Osteoarthritis%20occurs%20when%20the%20cartilage,bone%20will%20rub%20on%20bone... [Accessed 5 July 2000].

[4] J. Michalak, N. Troje, J. Fischer, P. Vollmar, T. Heidenreich and D. Schulte, "Embodiment of Sadness and Depression—Gait Patterns Associated With Dysphoric Mood," *Psychomatic Medicine: Journal of Behavioural Medicine,* vol. 71, no. 5, p. June, 2009.

[5] "Understanding Phases of the Gait Cycle," ProtoKinetics, 28 November 2018. [Online]. Available: https://www.protokinetics.com/understanding-phases-of-the-gait-cycle/. [Accessed 5 July 2022].

[6] P. L. S. B. T. R. S. R. S. Amit Gupta, "Knee Flexion/Extension Angle Measurement for Gait Analysis Using Machine Learning Solution "MediaPipe Pose" and Its Comparison with Kinovea," *IOP Conference Series: Materials Science and Engineering,* 2022.

[7] F. &. E. M. &. S. Z. &. J. M. Vaverka, "System of gait analysis based on ground reaction force assessment," *Acta Gymnica.,* 2015.

[8] B. G.-Z. A. M.-Z. Alvaro Muro-de-la-Herran, "Gait Analysis Methods: An Overview of Wearable and Non-Wearable Systems, Highlighting Clinical Applications," *Sensors,* vol. 14, no. 2, 2014.

[9] M. Haghighat, "Challenges of force plate and infrared camera-based gait analysis for cost-effective use," *IEEE Transactions on Biomedical Engineering,* vol. 68, no. 4, 2021.

[10] A. M. Howell, "Insole-based gait analysis," The University of Utah ProQuest Dissertations Publishing, 2012.

[11] A. M. Howell, T. Kobayashi, H. A. Hayes, K. B. Foreman and S. J. M. Bamberg, "Kinetic Gait Analysis Using a Low-Cost Insole," *IEEE Transactions on Biomedical Engineering* , vol. 60, no. 12, 2013.

[12] A. Rezaei, A. Ejupi, M. Gholami, A. Ferrone and C. Menon, "Preliminary Investigation of Textile-Based Strain Sensors for the Detection of Human Gait Phases Using Machine Learning," in *IEEE International Conference on Biomedical Robotics and Biomechatronics (Biorob)*, Enschede, Netherlands, 2018.

[13] T.-H. H. S. Y. Jianfu Yang1, "Machine Learning Based Adaptive Gait Phase Estimation," in *Proceedings of the 2019 Design of Medical Devices Conference*, Minneapolis, 2019.

[14] J. Pineda-Gutierrez, L. Miró-Amarante, M. Hernandez-Velazquez and F. Sivianes-Castillo, "Designing a Wearable Device for Step Analyzing," *IEEE 32nd International Symposium on Computer-Based Medical Systems (CBMS),* 2019.

[15] A. Q. Caddle, K. Gilhooly, L. Govoni and M. L. Sivak, "Design of Patient-Specific Ankle-Foot Orthotics," *Capstone Design Program: Mechanical Engineering.*

[16] H. Ma, Y. Min, F. Wu, X. Gao, X. Ma, J. Yao, C. Ma and X. Li, "The Influence of Walking Height and Width on the Gait," *Journal of Healthcare Engineering,* vol. 24, p. June, 2021.

[17] M. E. a. M. R. Espandar, "Biomechanical Factors Contributing to Higher Ground Reaction Force During Toe-Off Phase of the Gait Cycle," *IEEE Journal of Translational Engineering in Health and Medicine,* vol. 9, pp. 1--8, 2021.

[18] S. K. S. a. A. K. A. K. Choudhary, "Analysis of Biomechanical Factors Affecting Ground Reaction Force During the Gait Cycle," *Proceedings of the 2020 IEEE 7th Uttar Pradesh Section International Conference on Electrical, Electronics and Computer Engineering (UPCON),* pp. 236-240, 2020.

[19] "https://learn.adafruit.com," adafruit, 29 July 2012. [Online]. Available: https://learn.adafruit.com/force-sensitive-resistor-fsr.

[20] I. Electronics, "https://www.interlinkelectronics.com/," [Online]. Available: https://www.interlinkelectronics.com/fsr-402.

[21] J.-L. Chen, Y.-N. Dai, N. S. Grimaldi, J.-J. Lin, B.-Y. Hu, Y.-F. Wu and S. Gao, "Plantar Pressure-Based Insole Gait Monitoring Techniques for Diseases Monitoring and Analysis: A Review," *Advanced Materials Technologies,* vol. 7, no. 17, 2021.

[22] J. Zou, L. Yu and Z. Ma, "Study on prevention of symptoms based on plantar pressure array typesensor composite system," *Journal of Physics: Conference Series,* vol. 2390, 2022.

[23] W. Lee, S.-H. Hong and H.-W. Oh, "Characterization of Elastic Polymer-Based Smart Insole and a Simple Foot Plantar Pressure Visualization Method Using 16 Electrodes," *Sensors,* vol. 19, no. 1, 2018.

[24] R. Ramalingame, Z. Hu, C. Gerlach and O. Kanoun, "Shoe insole with MWCNT-PDMS-composite sensors for pressure monitoring," *SENSORS,* 2017.

[25] A. N. Fleury, "Characterization and Clinical Applications of Textile-Based Sensors in Rehabilitation," University of Toronto, Toronto, 2018.

[26] S. Li, "Numerical Investigation of Deformations of Materials with Embedded Systems and Human Foot Structure under Different Loading Conditions," Liverpool John Moores University , 2018.

[27] S. Li, "Experimental Testing and Numerical Investigation of Materials with Embedded Systems during Indentation and Complex Loading Conditions," Liverpool John Moores University., 2018.

[28] Ribeiro and C. Enes, "Design of a wearable active ankle-foot orthosis for both sides," University of Minho, 2021.

[29] "A Force-Sensing Insole to Quantify Impact Loading to the Foot," *Journal of Biomechanical Engineering,* vol. 141, no. 2, 2018.

[30] C. S. S. M. N. A. Senanayake, "Human Assisted Tools for Gait Analysis and Intelligent Gait Phase Detection," *Conference on Innovative Technologies in Intelligent Systems and Industrial Applications,* 2009.

[31] C. M. L. S.-M. K. M. I. a. C. H. K. Ji Su Park, "Gait Phase Detection Using Force Sensing Resistors," *IEEE SENSORS JOURNAL,* 2020.

[32] C. M. S. a. S. M. N. A. Senanayake, "Computational Intelligent Gait-Phase Detection System to Identify Pathological Gait," *IEEE TRANSACTIONS ON INFORMATION TECHNOLOGY IN BIOMEDICINE,* 2010.

[33] X. O. Z. L. a. H. Y. Shuo Ding, "Proportion-Based Fuzzy Gait Phase Detection Using the Smart Insole," *Sensors and Actuators A: Physical,* 2018.

[34] M. R. P. T. K. V. D. a. M. M. Ion P. I. Pappas, "A reliable gait phase detection system," *IEEE Transactions on Neural Systems and Rehabilitation Engineering,* 2021.

[35] M. E. H. C. A. K. S. A.-H. M. A. S. A. M. B. I. R. N. A.-E. Anas M. Tahir, "A Systematic Approach to the Design and Characterization of a Smart Insole for Detecting Vertical Ground Reaction Force (vGRF) in Gait Analysis," *MDPI,* 2020.

[36] J. Z. a. c. Y. W. a. c. Z. S. b. E. Z. a. c. Lie Yu a c, "Adaptive method for real-time gait phase detection based on ground contact forces," *MDPI,* 2019.

[37] J. F. *. H. a. B. Iván GonzálezORCID, "An Ambulatory System for Gait Monitoring Based on Wireless Sensorized Insoles," *MDPI,* 2015.

[38] H. Cho, "Design and Implementation of a Lightweight Smart Insole for Gait Analysis," *IEEE,* 2017.

[39] A. M. Howell, T. Kobayashi, H. A. Hayes, K. B. Foreman and S. J. M. Bamberg, "Kinetic Gait Analysis Using a Low-Cost Insole," *IEEE,* 2018.

[40] Y.-C. L. W. L. K. Z. Song-Hua Yan, "Gait phase detection by using a portable system and artificial neural network," *ELSEVIER,* 2021.

[41] W. N. I. &. T. K. Srivises, "Design of a smart shoe for reliable gait analysis using state transition theory.," *IEEE,* 2012.

[42] "Foot Bones," The Skeletal System, [Online]. Available: https://www.theskeletalsystem.net/foot-bones. [Accessed 8 May 2023].

[43] N. K. a. S. K. Kaur, "Design and Development of a Force Sensing Resistor (FSR) Based Circuit for Measuring Force," *International Conference on Electronics and Sustainable Communication Systems (ICESC),* pp. 1-5, 2020.

[44] R. MISCHIANTI, "DOIT ESP32 DEV KIT v1 high resolution pinout and specs," 13 January 2023. [Online]. Available: https://www.mischianti.org/2021/02/17/doit-esp32-dev-kit-v1-high-resolution-pinout-and-specs/. [Accessed 4 May 2023].

[45] T. T. H. Duong, D. Uher, J. Montes and D. Zanotto, "Ecological Validation of Machine Learning Models for Spatiotemporal Gait Analysis in Free-Living Environments Using Instrumented Insoles," *IEEE Robotics and Automation Letters,* vol. 7, no. 4, 2022.

[46] M. Lueken, L. Mueller, M. G. Decker, C. Bollheimer, S. Leonhardt and C. Ngo, "Evaluation and Application of a Customizable Wireless Platform: A Body Sensor Network for Unobtrusive Gait Analysis in Everyday Life," *MDPI,* vol. 20, 2020.

[47] P. Khera and N. Kumar, "Role of machine learning in gait analysis: a review," *Journal of medical engineering and technology,* vol. 44, no. 8, 2020.

[48] S. M. Alfayeed and B. S. Saini, "Human Gait Analysis Using Machine Learning: A Review," *International Conference on Computational Intelligence and Knowledge Economy (ICCIKE),* 2021.

[49] "What is machine learning?," IBM, [Online]. Available: https://www.ibm.com/topics/machine-learning. [Accessed 8 May 20203].

[50] J. M. Carew, "reinforcement learning," Tech Target, [Online]. Available: https://www.techtarget.com/searchenterpriseai/definition/reinforcement-learning#:~:text=Reinforcement%20learning%20is%20a%20machine,learn%20through%20trial%20and%20error.. [Accessed 8 May 2023].

[51] "https://www.techtarget.com/searchenterpriseai/definition/reinforcement-learning#:~:text=Reinforcement%20learning%20is%20a%20machine,learn%20through%20trial%20and%20error.," AI and Machine learning, 10 March 2023. [Online]. Available: https://www.simplilearn.com/tutorials/machine-learning-tutorial/types-of-machine-learning. [Accessed 8 May 2023].

[52] D. Soni, "Supervised vs. Unsupervised Learning," Towards Data Science, 22 March 2018. [Online]. Available: https://towardsdatascience.com/supervised-vs-unsupervised-learning-14f68e32ea8d#:~:text=Supervised%20learning%20is%20typically%20done,input%20to%20a%20continuous%20output.. [Accessed 8 May 2023].

[53] J. Terra, "Regression vs. Classification in Machine Learning for Beginners," AI and Machine learning, 22 February 2023. [Online]. Available: https://www.simplilearn.com/regression-vs-classification-in-machine-learning-article#:~:text=Regression%20algorithms%20solve%20regression%20problems,cancer%20cells%2C%20and%20speech%20recognition.. [Accessed 8 May 2023].

[54] S. Gupta, "Regression vs. Classification in Machine Learning: What's the Difference?," Springboard, 6 October 2021. [Online]. Available: https://www.springboard.com/blog/data-science/regression-vs-classification/. [Accessed 8 May 2023].

[55] J. Brownlee, "Difference Between Classification and Regression in Machine Learning," Machine Learning Mastert, 22 May 2019. [Online]. Available: https://machinelearningmastery.com/classification-versus-regression-in-machine-learning/. [Accessed 8 May 2023].

[56] "Classification Algorithm in Machine Learning," Javapoint, [Online]. Available: https://www.javatpoint.com/classification-algorithm-in-machine-learning. [Accessed 8 May 2023].

[57] D. Gong, "Top 6 Machine Learning Algorithms for Classification," Towards Data Science, 23 February 2022. [Online]. Available: https://towardsdatascience.com/top-machine-learning-algorithms-for-classification-2197870ff501. [Accessed 8 May 2023].

[58] "7 Types of Classification Algorithms in Machine Learning," ProjectPro, 2 May 2023. [Online]. Available: https://www.projectpro.io/article/7-types-of-classification-algorithms-in-machine-learning/435. [Accessed 8 May 2023].

[59] "Classification in Machine Learning: What it is & Classification Models," AI and Machine learning, 29 March 2023. [Online]. Available: https://www.simplilearn.com/tutorials/machine-learning-tutorial/classification-in-machine-learning. [Accessed 8 May 2023].

[60] J. Brownlee, "What is the Difference Between Test and Validation Datasets?," Machine learning process, 14 August 2020. [Online]. Available: https://machinelearningmastery.com/difference-test-validation-datasets/. [Accessed 14 May 2023].

[61] J. Doe, "Why do different classification algorithms of supervised machine learning give different accuracy score?," AI News, 1 May 2023. [Online]. [Accessed 10 May 2023].

[62] J. Brownlee, "What is a Confusion Matrix in Machine Learning," Learning mastery, 15 August 2020. [Online]. Available: https://machinelearningmastery.com/confusion-matrix-machine-learning/#:~:text=A%20confusion%20matrix%20is%20a,two%20classes%20in%20your%20dataset.. [Accessed 8 May 2020].

[63] "Gait Analysis: Technology and Clinical Applications," 6 June 2015. [Online]. Available: https://clinicalgate.com/gait-analysis-technology-and-clinical-applications/. [Accessed 7 July 2022].

[64] "www.kistler.com," Kistler Instrument Corp, [Online]. Available: https://www.kistler.com/en/glossary/term/ground-reaction-force-grf/.

[65] S. K. Mitra, "Digital signal processing: A review journal," *IEEE Transactions on Signal Processing,* vol. 48, no. 6, 2000.

[66] E. Burns, "machine learning," Techn Target, [Online]. Available: https://www.techtarget.com/searchenterpriseai/definition/machine-learning-ML. [Accessed 8 May 2023].

[67] A. K. Sharma, "Predicting Gait Parameters of Leg Movement with sEMG and Accelerometer Using CatBoost Machine Learning," *electronics,* vol. 13, no. 9, p. 1791, 2024.

[68] S. M. Alfayeed, "Human Gait Analysis Using Machine Learning: A Review," in *2021 International Conference on Computational Intelligence and Knowledge Economy (ICCIKE)*, Dubai, 2024.

[69] S. S. P. B. Slavomir KARDOS, "Sensor Array for Evaluation of Gait Cycle," vol. 14, no. 4, 2019.

IOP Conf. Series: Materials Science and Engineering **1314** (2024) 012009 doi:10.1088/1757-899X/1314/1/012009

Comparative Study of Different Erosion Models in Francis Runner Blade Using OpenFOAM

Suprim Shrestha[1,2,3], Pawan Lal Bijukchhe[2], Sailesh Chitrakar[1,2,3], Bhola Thapa[2,3], Zhongdong Qian[1,3] and Zhiwei Guo[1,3]

[1]State Key Laboratory of Water Resources Engineering and Management, Wuhan University, China
[2] Turbine Testing Lab, Kathmandu University, Nepal
[3]Hydropower Research Centre for Himalayan Region

E-mail: suprim78@gmail.com and sailesh@ku.edu.np

Abstract. Francis turbine serves as a prominent hydro turbine choice for harnessing water-based energy. However, the presence of sediment particles in river environments poses a considerable challenge for constructing hydropower facilities along sediment-laden rivers. This challenge is particularly pronounced in scenarios involving high- and medium-head hydroelectric power installations. Erosion emerges as a significant concern in such contexts due to the abrasive impact of sediment particles on hydro turbines, which can have adverse effect on its performance, flow dynamics, potentially leading to operational failures. Erosion model is one of the methods used for predicting the erosion rate in hydro turbines. State-of-the-art modelling and simulation using computational fluid dynamics (CFD) are widely adopted methods for predicting erosion. In this study, various commonly used erosion models were implemented and followed by a comparative analysis of the erosion rates predicted by each model. The study utilizes a model runner of Jhimruk Hydropower Plant using OpenFOAM for conducting numerical simulation. The results show variations among the models in identifying areas with the erosion rates due to factors considered by each model. These insights provide a better understanding of how erosion occurs and help in developing strategies for optimizing hydro turbine performance in sediment-laden environments.

1. Introduction

Sedimentation is a worldwide issue that has been observed in the rivers of various regions, including the Himalayan region in China, Asia, and the Andes in South America (1). The Himalayan Mountain range, in Nepal has a potential for generation due to the ample water resources originating from the snow-covered peaks and glaciers. However, power plants that rely on river flow face a challenge in the form of sediment erosion, which greatly affects the utilization of hydropower. Most rivers in the Himalayan region contain around 60% to 80% particles with hardness levels exceeding 6 on the Moh's scale (2). These hard particles cause wear and damage to turbine components leading to reduced turbine efficiency and a shorter lifespan, for operation. Erosion damage, to turbines is a process that is affected by various factors, such, as the characteristics of the sediment particles (their shape, size and hardness) the concentration of particles the speed at which particles move and the angles at which they collide with surfaces. Additionally, the erosive damage is also affected by the properties of the materials used in turbine manufacturing (3). The intricate flow pattern associated with the Francis turbine causes sediment particles to strike the turbine surface in an unpredictable fashion. Numerous studies have shown that erosion primarily occurs at the entry and exit points of runner blades in turbines. This is mainly due to the high-velocity of water flow in these regions. The notable erosion rate in Francis turbines is mainly a result of the high relative speed at the runner's exit. In real operational scenarios, guide vanes adjustment is the best way to change how the turbine works. However, under non-standard conditions, the flow can separate, leading to the creation of swirling patterns called inter-blade vortexes. These vortices channel particles to a specific area where they gather, causing increased accumulation and worsening of erosion. This implies that inter-blade vortexes play an important role in causing severe wear and tear at the blade exit (4). Figure 1 shows erosion in runner blades and a guide vane of Francis turbine.

Figure 1. Erosion in Francis runner blades and a guide vane

1.1 Erosion model

Erosion models are crucial for designing turbine components, planning sediment settling basins, and optimizing the operations of hydropower plants in rivers with high sand content. Typically, these models rely on the analysis of the behaviour of individual particles to develop effective erosion models. Some erosion models in hydraulic machinery have been established using empirical and statistical relationships derived from experiments and field measurements. However, it is important to note that only a limited number of these erosion models have been rigorously tested and validated for their reliability and accuracy (5).

The simplest form of erosion equation can be written as Erosion = f (operating condition, properties of particles, and properties of base material).

In general, erosion in mechanical systems and Francis turbines is a complex phenomenon, where multiple factors intervene; however, operating conditions, particle properties, and base material used are the most important. In most cases, the erosion rate will be dependent on the fluid velocity or speed; higher velocities usually result in increased erosion because of the high particle impact energy. The degree of hardness is another factor; harder materials are prone to performing in better resistance to erosion. Moreover, particle size and fluid concentration are also important, as larger and more concentrated particles cause significantly more wear than smaller particles.

General models of erosion have been developed to predict the erosion rate. The most frequently cited expression for erosion, as identified by Truscott (6), is

$$Erosion \propto (Velocity)^m \tag{1}$$

Some of the models have been specifically used to predict erosion for hydraulic turbines. Thapa (7) undertook laboratory testing to determine the erosion rates for a range of materials used for turbines, under a variety of different conditions. His results for 16Cr5Ni the most common material used in turbines yielded the empirical relation given below, which predicts the erosion rate.

$$y = 6E - 5x^{3.13} \ \text{[mg/kg]} \tag{2}$$

Here, x (m/s) represents eroding particles velocity impacting at a 45-degree angle, y denotes the material loss in milligrams per kilogram of particles hitting the surface.

Bajracharya(8) has formulated the following empirical relations to estimate the erosion rate.

$$E_r = a.(size)^b \text{ [mm/year]} \quad (3)$$

Where a and b are empirical constants which value depends according to quartz content percent.

The parameters that influence turbine erosion were correlated by Tsugio (9) using the erosion data obtained from different hydropower plants. The relation correlating erosion in turbines suggested by him is as follows:

$$W = \beta.C^x.a^y.k_1.k_2.k_3.V^m \quad \text{[mm/year]} \quad (4)$$

Here V denotes the flow velocity. k_1, k_2 are coefficient of shape and hardness of sand particles. k_3 is the abrasion resistance coefficient of the material. The values x and y are the exponents for concentration and size coefficients.

The erosion models also have their own limitations. While the general erosion models are of wide application and provide insight into the processes of erosions in general, they largely oversimplify the complex particle-material interactions. They rely too much on site-specific constants which vary to a great extent hence compromising the model's precision when used under different conditions. Even the model developed by Tsuguo, which is based on a huge amount of data, has the same problems because the empirical coefficients involved are site-specific and hence of no significant use in making high-quality general predictions. There is little information about interactions between the particles and fine details about the material properties in these models, hence less reliability. In particular, the empirical models by Bajracharya targeted predictions for the erosion rates and efficiency losses; they are deficient in including such critical factors as particle concentration, shape, and hardness, and the results are expressed through proportionality functions misleading without accurate constants.

1.2 Jhimruk Hydro-Electric Centre

The Jhimruk Hydro-Electric Centre is a run-of-river hydropower plant with a capacity of 12 MW, featuring three Francis turbines, each operating at 1000 rpm. This facility utilizes a total head of 201.5 meters and a flow rate of 2.35 cubic meters per second. Since its inception, the plant has encountered a significant issue with erosion. While some measures have been implemented to mitigate erosion, the primary solution thus far has been to conduct maintenance and repairs on the eroded components to keep the plant operational. An analysis of sediment in the river revealed that the majority of particles entering the runner have diameters less than 100 microns, with an average size of 25 microns (10). The sediment concentration in the river varies between 500 parts per million (ppm) and 57,000 ppm (11). Notably, the river contains a high proportion of hard minerals, with quartz being the predominant mineral (12).

The primary objective is to conduct a comparative study of different erosion models (Finnie, DNV, Nandakumar, and Arabnejad) applied to the Francis runner blade using OpenFOAM, an open-source CFD package. The study aims to predict erosion rates in a Francis turbine runner blade operating under sediment-laden conditions, identifying areas with significant erosion and understanding the influence of factors such as impact angle, particle velocity, and material properties.

The study's focus on a single operating point limits its ability to capture the full range of operational conditions.

2. Mathematical Model

Computational fluid dynamics (CFD) plays a pivotal role in modern engineering, primarily focusing on effectively examining the behavior of a studied physical phenomenon or effect. This examination is accomplished by solving the differential equations that govern the preservation of mass, momentum, and energy in moving fluids. CFD methods are rooted in numerically computing the flows of liquids and gases by solving the Navier-Stokes and continuity equations, which represent the most general scenarios of fluid motion. The equations utilized for this numerical analysis of fluid flow adhere to the Reynolds Averaged Navier Stokes (RANS) equation and employ turbulence models based on the concept of eddy viscosity (13).

While the utility of CFD varies by industry, it finds significant application in hydraulic engineering, particularly in hydropower. Here, its primary use revolves around understanding fluid flow behaviour, the interaction between solids and fluids (solid-fluid interaction or FSI), and optimizing design proposals based on the obtained results. Numerous commercial software packages and open-source tools are employed for conducting numerical studies. In this study, CFD simulations were performed using OpenFOAM.

The governing equations RANS for continuity and momentum for RANS are given in equation 5-9.

$$\frac{\partial p}{\partial t} + \nabla \cdot (PV) = 0 : \text{Conservation of mass} \tag{5}$$

$$\rho \frac{DV}{Dt} = \rho g + \nabla . \tau' ij - \nabla p : \text{Conservation of momentum} \tag{6}$$

$$\rho \frac{Dh}{Dt} = \frac{Dp}{Dt} + \nabla(k \nabla T) + \emptyset : \text{Conservation of energy} \tag{7}$$

Incompressible fluid flow is the main subject of the Computational Fluid Dynamics (CFD) numerical analysis of the Francis turbine. Equations 8 and 9, which express the conservation of mass and momentum, are made simpler as a result. Furthermore, the analysis regards the influence of temperature on the flow as insignificant. As a result, during the analysis, energy conservation is ignored.

$$\nabla . V = 0 \tag{8}$$

$$\rho \frac{DV}{Dt} = \rho g + \mu \nabla^2 V - \nabla p \tag{9}$$

The two most commonly used two-equation models in turbulence modelling are the standard k-ω model with or without a wall function and the k-Ω model. One selects one of them according to the flow configuration that one intends to investigate. Typically, the k-ω model is usually more robust in terms of obtaining convergence, especially in regions of the flow with relatively basic geometry and with rapid exciting transition not so critical. However, limitation exists in the reliable performance of this model with some phenomena, like flow with a free surface and effects of boundary layers. It is numerically known to converge quickly with a limitation in these domains. The k-Ω turbulence model operates relatively well in wall-bounded regions and provides quite realistic results in them. However, it has a relatively slower convergence rate compared to the k-ω model and often requires more computational capacity. An excellent compromise is the SST (Shear Stress Transport) model, which merges the benefits of the k- Ω model in the near-wall regions with the k- ω model in the open flow sections. To simulate complex flow features, the advantages provided by this model yield a well-balanced procedure regarding both accuracy and computational cost-effectiveness (14). SST turbulence model was used for this study.

IOP Conf. Series: Materials Science and Engineering **1314** (2024) 012009 doi:10.1088/1757-899X/1314/1/012009

OpenFOAM is an open-access CFD tool for simulating heat transfer, fluid flow, and other related processes. The software is pretty useful in engineering and scientific research as it provides a flexible framework with various solvers and libraries. Being open-source code, users can adapt and tailor the code to meet specific needs. In this work, the Lagrangian Intermediate Library has been combined with SRFSimpleFOAM, a steady-state solver which models non-Newtonian fluid flow in a rotating frame, ensuring that the fluid is incompressible and turbulent. This solver defines the velocity field as relative velocity and is especially helpful in rotating systems. The Lagrangian Intermediate Library provides templated functions, models, and forces to compute particle variables conditioned on the resolved carrier phase. Applicable to cases without chemical reaction and temperature effects, the class "Kinematic" will simulate the behaviour of particles in fluid flow.

2.1 Particle Motion Model

A fluid containing motion of groups of particles can be simulated in two ways: the Eulerian-Eulerian way and the Eulerian-Lagrangian. In the former approach, the particles are modelled as a continuous medium while equations of conservation, as in the case of single-phase flow, describe their behaviour. This method works well where there is a high concentration in particles in the fluid. On the other hand, the E-L method involves basically solving equations for the fluid phase and Newton's equations for the particle phase in order to track how particles move. It finds a better application when there is a low concentration of particles in fluid (15). The motion equation of the particles is expressed by:

$$F_p = m_p \frac{dV_p}{dt} = F_D + F_p + F_g + F_A \tag{10}$$

F_D denotes drag force acting on the particle, F_p is the pressure gradient force, F_g indicates gravity and buoyancy force and F_A is the added mass force.

Drag force is calculated as:

$$F_D = m_p \frac{18\mu}{\rho_p d_p^2} \frac{C_D Re(Re)}{24} (u - u_p) \tag{11}$$

C_D, drag coefficient given by the following equation:

$$C_D = \begin{cases} \frac{24}{Re_p}\left(1 + 0.15 Re_p^{0.687}\right) & if\ Re_p \leq 1000 \\ 0.44 & if\ Re_p > 1000 \end{cases} \tag{12}$$

The gravity and buoyancy force are:

$$F_g = m_p g \left(1 + \frac{\rho}{\rho_p}\right) \tag{13}$$

The Pressure gradient force is:

$$F_p = \frac{1}{6} \pi d_p^3 \nabla P \tag{14}$$

And the added mass force:

$$F_A = \frac{1}{12}\pi d_p^3 \rho_p \frac{dV_p}{dt} \tag{15}$$

3. Erosion Modelling

The strong language feature of C++ templates is frequently utilized by the Kinematic library. To calculate erosion precisely within the Kinematic class, it makes use of a template function. When particles or computational units strike certain walls in the target geometry where erosion analysis is necessary, it records information about impacts, including speed and angle. The captured data is then utilized to determine erosion at each distinct border face. All of the border faces' combined data is recorded in a file for future study. By adding the erosion that each particle causes when it reaches the prescribed borders, the erosion at each boundary face is computed (16).

3.1 Implemented Erosion Model:

In this work, besides the inbuilt Finnie erosion model, three other models (DNV, Arabnejad, and Nandakumar) were implemented in OpenFOAM. The erosion results were compared by coupling these models with an incompressible solver and recompiling the code. To facilitate this, a "src/lagrangian" folder was created in the user directory, and necessary files were copied there. The "particleErosion" template in OpenFOAM, which includes "ParticleErosion.H" and "ParticleErosion.C" files, was used to calculate erosion on specified patches. "ParticleErosion.H" defined interfaces, data structures, and function prototypes, while "ParticleErosion.C" contained the actual source code for simulating material erosion from particle collisions. This setup allowed for the accurate simulation and comparison of erosion patterns across different models. The integration of these models required precise modifications and recompilations to ensure that the particle behavior and erosion calculations were accurately represented. This comprehensive approach enabled a detailed analysis of how different erosion models predict material loss due to sediment impact, providing valuable insights for optimizing turbine design and improving durability in sediment-laden environments.

3.1.1 Finnie Erosion Model:

The Finnie erosion model includes two equations based on the impingement angle. These formulas estimate the volume of material removed by an individual abrasive particle with mass $m$and velocity V. The model uses two critical experimental constants: the ratio of the depth of contact to the depth of cut and the ratio of vertical to horizontal force components. It also incorporates plastic flow stress and impingement angle. The Finnie model is accurate for low impingement angles, common in ductile metals, but initially predicted no erosion at normal incidence. To address this, Finnie modified the model to account for erosion at normal incidence angles, resulting in the equations given in (16) and (17).

$$Q = \frac{mV^2}{p\varphi K}(sin2\alpha - \frac{6}{k}\sin^2\alpha) \text{ If } \tan\alpha \leq \frac{K}{6} \tag{16}$$

$$Q = \frac{mV^2}{p\varphi K}(\frac{Kcos^2\alpha}{6}) \text{ If } \tan\alpha \geq \frac{K}{6} \tag{17}$$

where p denotes plastic flow stress of the material, α indicates angle of impingement, φ represents the ratio of depth of contact to the depth of cut and K represents the ratio of vertical to horizontal force components on the particle.

3.1.2 DNV Erosion Model

Det Norse Veritas, (DNV) (18) developed an erosion model specifically for straight pipes, elbows, plugged tees, welded joints, and reducers. This model was created using extensive experimental data and numerical predictions. It is expressed as follows:

$$ER = CF(\theta)V_p^n \tag{18}$$

$$F(\theta) = \sum_{i=1}^{8}(-1)^{i+1}A_i\theta^i \tag{19}$$

Table 1. Coefficients corresponding to DNV model.

A1	A2	A3	A4	A5	A6	A7	A8
9.370	-42.295	110.864	-175.804	170.137	-98.398	31.211	-4.170

where ER denotes the erosion rate of the target, defined as the wall mass loss per unit area and per unit time. F (θ) represents the impact angle function, C is a constant (for steel pipes, C = 2.0 x 10^{-9}), n is the velocity exponent (for steel pipes, n = 2.6). The values of Ai are listed in Table 1.

3.1.3 Nandakumar Erosion Model

Nandakumar et al. (19) based their erosion model on the work by Bitter and Finnie. This model predicts the volume eroded by further particle collisions on a surface and divides erosion into cutting and deformation wear. Equation 20, used in the simulation, shows how the model developed as

$$\Delta Q = Cm\rho_p^{0.15}(V_0 sin\theta)^{2.3}Dm^{1.1875}d_p^{-0.0625}V_0^{2.375}(cos\theta)^2(sin\theta)^{0.375} \tag{20}$$

where C and D are empirical constant, whose values are considered $7.5 \cdot 10^{-4}$ and 0.082, respectively, whereas m represents the particle mass, $\rho_p$is particle density, θ is the impact angle, V_0 is the particle velocity, and d_p is the diameter of particle.

3.1.4 Arabnejad Erosion Model

A new erosion equation by Arabnejad et al. (20) considers the erosion generated by particle collisions as caused by two processes: cutting and deformation. Equation used within the model is:

$$ER = ER_c + ER_D \tag{21}$$

$$ER_C = \begin{cases} C_1F_s\frac{V_p^n\sin(\theta)[2Kcos(\theta)-\sin(\theta)]}{2K^2} & \theta < \tan^{-1}(K) \\ C_1F_s\frac{V_p^n\cos^2(\theta)}{2} & \theta > \tan^{-1}(K) \end{cases} \tag{22}$$

$$ER_D = C_2F_s\left(V_p\sin(\theta) - V_{tsh}\right)^2 \tag{23}$$

Table 2. Empirical constants for erosion equation

Material	C_1	C_2	K	V_{tsh}
Carbon steel 1018	5.90E-08	4.25E-08	0.5	5.5
Carbon steel 4130	4.94E-08	3.02E-08	0.4	3.0
Stainless steel 316	4.58E-08	5.56E-08	0.4	5.8
Stainless steel 2205	3.92E-08	2.30E-08	0.4	2.3
13 chrome duplex	4.11E-08	3.09E-08	0.5	5.1
Inconel 625	4.58E-08	4.22E-08	0.4	5.5
Aluminium alloy 6061	3.96E-08	3.38E-08	0.4	7.3

where is particle velocity, θ is impact angle, Fs is particle shape factor and n are velocity exponent used as 2.41. C_1, C_2, K and V_{tsh} are empirical constants which value varies according to the material used and are listed in Table 2.

4. Numerical Setup

4.1 Geometry Model and Mesh

For the numerical analysis, the geometry and mesh were created using ANSYS TurboGrid. To reduce computational effort while maintaining precision, only a single passage runner fluid domain of the turbine was simulated. This domain includes an inlet, hub, shroud, single runner blade, and an outlet domain. The computational domain and the names of the patches are shown in Figure 3. ANSYS TurboGrid was used to generate a hexahedral structured mesh, with the mesh statistics presented in Table 3.

Table 3. Mesh Statistics

Component	Mesh Type	No. of Elements	No. of Nodes
Single Passage Domain	Hexahedral	242100	264895

IOP Conf. Series: Materials Science and Engineering **1314** (2024) 012009 doi:10.1088/1757-899X/1314/1/012009

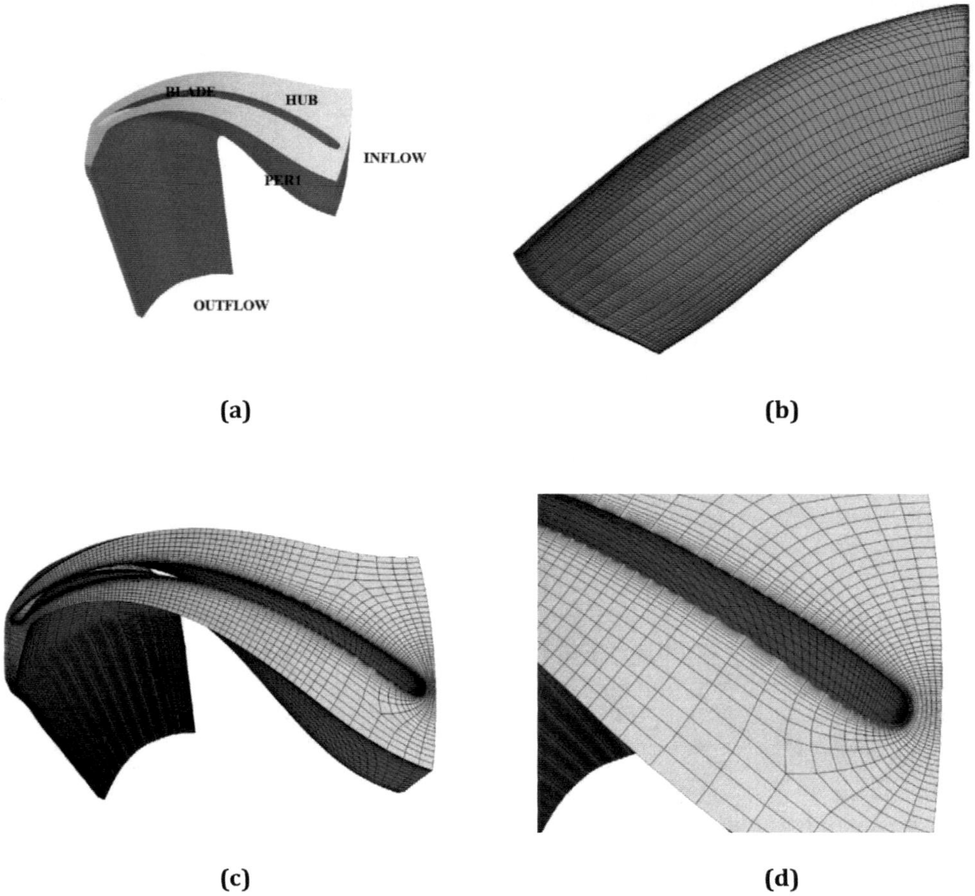

Figure 2. (a) Single passage domain, (b) Blade mesh, (c) Domain Mesh, (d) Zoom view of the Mesh around blade area

4.2 Boundary Conditions

For the simulation done with the SRFSimpleFoam solver, inlet conditions were provided for the relative velocity, using swirlInletVelocity type. This had a radial velocity of -8.49 m/s and tangential velocity of 11.14 m/s. Velocity at the outlet boundary was set with a zeroGradient condition, and at the outflow pressure it is constant, with 1 atmosphere. The hub, shroud, and blade, they were defined as walls, and the noSlip type condition was applied to these specific patches. The details of boundary condition are presented on Table 4. And details of turbine specification is given in Table 5.

Table 4. Boundary Conditions

Patches	Urel	p	K	Omega
Inflow	swirlInletVelocity	zeroGradient	Fixed value uniform 0.289	Fixed value uniform 35
Outflow	zeroGradient	Fixed value uniform 101.32	inletOutlet	inletOutlet
Walls (hub, shroud, blade)	noSlip	zeroGradient	kqRWallFunction	omegaWallFunction

Table 5. Turbine Specifications

Parameter	Value
Head (m)	201.5
Flow (m3/s)	2.35
Power (MW)	4.2
Runner inlet diameter (mm)	890
Runner speed (rpm)	1000
No. of blades	17
No. of guide vanes	20

4.3 Sediment Injection

The OpenFOAM simulation uses the sediment properties given in Table 6. All the parameters of particles are defined in the "kinematicCloudProperties" file under the "constant" folder in the OpenFOAM case file directory. This file is used for defining common properties for all kinetic particles, forces acting on the particles, injection models, and other main parameters needed for erosion calculation. In OpenFOAM, erosion is quantified as a volume of materials removed from the system under the term "kinematicCloudQ."

Table 6. Sediment Properties

Characteristics	Value
Density (kg/m^3)	2650
Size (microns)	250
Concentration (ppm)	1400

IOP Conf. Series: Materials Science and Engineering **1314** (2024) 012009 doi:10.1088/1757-899X/1314/1/012009

5. Results and Discussion

5.1 Erosion wear in runner blade

Comparing the erosion patterns on the runner blade, obtained with the various erosion models, delivers some interesting results. Although for all the models, there is a general trend of maximum erosion on the pressure side and further up toward the trailing edge of the blade, this can be explained by the considerably higher relative velocity in the outlet region of the runner. However, the Nandakumar and Arabnejad models have a difference that has distinctly appeared. They plan a more marked pattern of erosion on the suction side of the blade, which is a result not quite in agreement with the literature.

Figure 3, showing this result from the Finnie Model, further reiterates this phenomenon. Actually, maximum erosion may be seen on the pressure side near the trailing edge, while a bit of erosion is also found around the leading edge on the suction side. This can mainly depend on the high acceleration at the inlet side of the blade.

(a) **(b)**

Figure 3. Volume of Material Eroded from Finnie Model (a)Suction side (b) Pressure side

Now turning to Figure 4, which gives the result obtained using the DNV model, one can realize that both the Finnie model and DNV model erosion trends are similar. They have almost a similar trend: maximum erosion found at the outlet region of the runner blades, and the leading edge on the suction side experiences a very minimal erosion. It's interesting to note that the DNV model brings out a slightly lower erosion value in the leading-edge area compared to the Finnie Model.

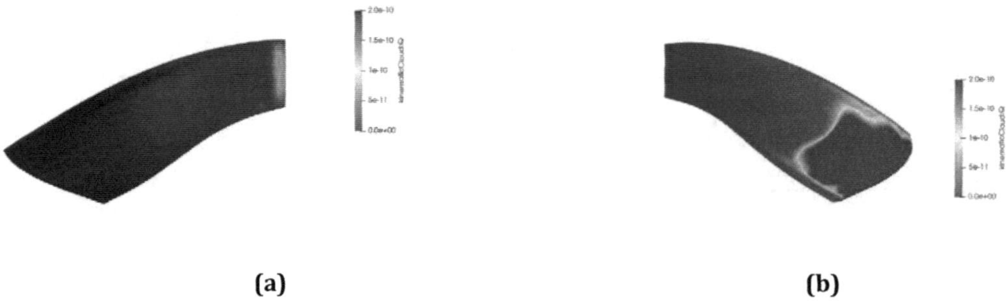

(a) **(b)**

Figure 4. Volume of Material Eroded from DNV Model (a)Suction side (b) Pressure side

Figure 5 shows the results of the NandaKumar Erosion Model. As one can clearly see in this plot, there are distinct red marks at both sides of the runner blades, which correspond to maximum erosion. However, one notable deviation from published literature is that high erosion on the suction side has been very prominent from the leading edge to the trailing edge. In comparison, it displays a greater area of erosion closer to the trailing edge on the pressure side than the other models.

(a) **(b)**

Figure 5. Volume of Material Eroded from Nandakumar Model (a)Suction side (b) Pressure side

Finally, Figure 6, which includes results from the Arabnejad model, illustrates another aspect of the erosion pattern. It clearly shows maximum erosion on both the pressure and suction sides of the blade. In addition, it indicates the great effect on the leading-edge region of the suction side, which is more than that obtained using the Finnie and DNV models. These characteristic patterns bring out an important understanding of the dynamics of erosion in relation to runner blades.

(a) **(b)**

Figure 6. Volume of Material Eroded from Arabnejad Model (a)Suction side (b) Pressure side

5.2 Erosion Wear Comparison with Actual Runner

This section assesses the erosion damage at Jhimruk Hydropower Plant and compares it with the results from CFD analysis in an attempt to find out which of the erosion models is close to the then-occurring actual erosion pattern existing in the power plant. It is already accepted that the numerical simulations may not exactly reproduce site operating conditions; however, they do predict similar erosion patterns. Figure 7 shows the erosion damage to the runner of the Jhimruk Hydropower Plant. It shows intensive erosion damage to the trailing edge outlet part and leading edge. What Figure 8 describes is related to the previous section results where the DNV model reported the predominant affected area of the manifold to be the runner outlet area. Inflow

IOP Conf. Series: Materials Science and Engineering **1314** (2024) 012009 doi:10.1088/1757-899X/1314/1/012009

impacts particulates to the leading edge which causes erosion. There is remarkable agreement between the numerical results of erosion and actual site erosion, especially in the areas that coincide with the predicted regions from the numerical analysis.

(a) **(b)**

Figure 7. Runner erosion from Jhimruk power plant (a) Outlet side, (b) Inlet side

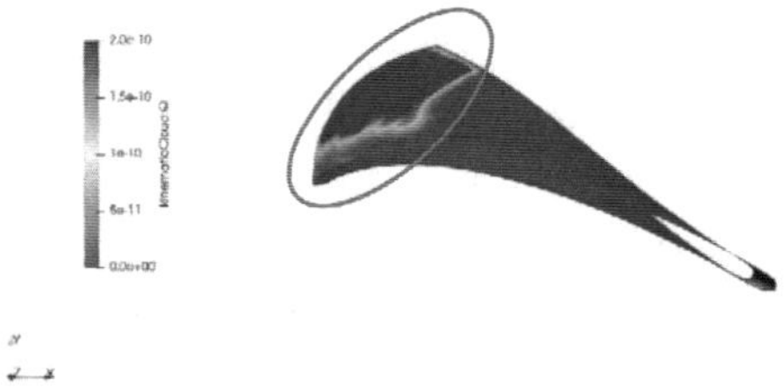

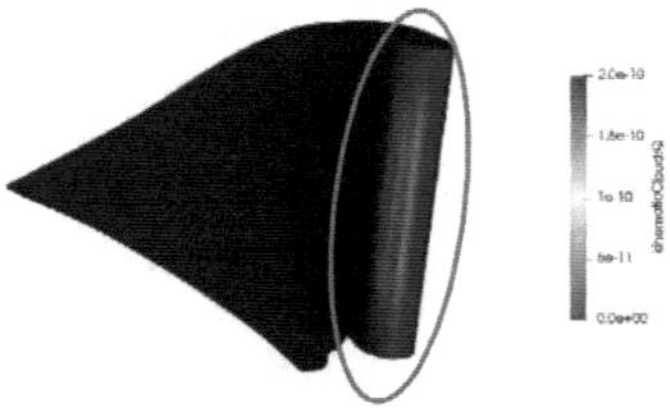

(a) **(b)**

Figure 8. Erosion predicted by DNV Model (a) Outlet side, (b) Inlet side

5.3 Variations Among the Applied Models

In this study, four different erosion models have been tested and compared for Francis runner blades. Each model has a different predicted erosion pattern. The differences among the above-mentioned erosion models used in this study come from their different sensitivities to impact angle, velocity, and particle properties.

The Finnie model is highly sensitive to the impact angle and predicts maximum erosion at low impingement angles. It initially showed no erosion at normal incidence, which was latter modified to cure this defect. In the Francis runner blade, the Finnie model showed maximum erosion near the trailing edge on the pressure side and minimum erosion on the suction side. This agrees with its prediction of higher erosion at low angles, which are normally experienced on the pressure side of the blade. Although it may predict erosion correctly for any high incidence angle and nonductile materials, this may not be the case. This may differ with other models that can account for a much wider range of incident angles and material types.

The Nandakumar model takes into consideration various factors, such as mass, density, and diameter of the particles, complex interaction between velocity and impact angle. In this model, the erosion area is larger, with clear distinct marks visible on both sides of the runner blades. It indicates maximum erosion from the leading edge to the trailing edge on the suction side, which is a very significant digression from all other models and published literature. In many cases, broader and more uniform erosion prediction could contradict the observation. This might be a case only when empirical constants are not well-calibrated to the conditions of the turbine.

For the Arabnejad model, this mainly covers cutting and deformation as the wear mechanisms. This dual consideration will, therefore, enable the model to predict erosion comprehensively, thereby capturing both high angle impacts, cutting, and continuous wear, deformation. The maximum erosion in both the Arabnejad model and at both pressure and suction sides of the blade occurs because it considers high angle impacts that result in cutting erosion and continuous impacts that result in deformation erosion. The high velocity, in addition to the impact angles being propitious for both mechanisms, results in considerable erosion at this location on the suction side leading edge. The leading edge on the suction side presents considerable erosion because particles impact this region at high velocities with poor angles that favour both cutting and deformation mechanisms. It is this nuanced behaviour that the detailed equations for ER_c and ER_D capture, resulting in these observed patterns. It is a complex model to have multiple empirical constants, which need to be accurately calibrated for different materials and conditions. Inaccurate determination of these constants would result in discrepancies between the predicted erosions. Moreover, being based on empirical data, the accuracy of the model is disproportionately dependent on the kind and range of data used for calibration. It may not predict erosion if the empirical data does not cover the full operational condition range.

The results obtained from the DNV erosion model showed very good agreement with the existing erosion patterns of the Francis turbine at the Jhimruk Hydropower Plant. The DNV model incorporates a detailed polynomial function $F(\theta)$ for very good capture of the impact angle dependence of erosion. This function is calibrated with multiple empirical coefficients Ai, which gives a nuanced capture of how different angles contribute to erosion. This all-inclusive angle function thus aids the model to very agreeably fit with the real observations since erosion in turbines is greatly dependent on the angles at which the particles impact these surfaces. Experimental data and numerical results form a large amount of background for the DNV model. It has been empirically calibrated for different conditions and materials; the constants used are very appropriate in making real-life erosion prediction. Very good agreement with material properties of the turbine components was obtained for model constants, especially for steel pipes.

5.4 Practical Implications and Recommendations for Turbine Optimization

These findings contribute to several practical implications, mainly in terms of turbine design optimization and adoption of erosion mitigation strategies. Based on the erosion patterns identified by the different models, redesigning turbine blades with thicker leading edges and more resistant materials in areas where erosion is more severe would improve durability and reduce frequency in maintenance. The CFD simulations can also be iteratively run to test different geometries of the blades with respect to minimizing areas of high impact velocity and optimizing flow paths so that there are fewer ongoing particle collisions with the blade surfaces. In this study, highly prone erosion areas are highlighted.

It is, therefore, possible to coat such areas with wear-resistant material like tungsten carbide or ceramic composite in order to elongate the life of the blades. The efficacies of the materials can be tested under simulated sediment-laden conditions. Knowing the response of erosion to various flow rates and angles can help in setting operational parameters so that performance could be optimized with reduced wear. For example, reducing flow rates during high sediment periods can decrease the erosion rate.

IOP Conf. Series: Materials Science and Engineering **1314** (2024) 012009 doi:10.1088/1757-899X/1314/1/012009

6. Conclusion

The aim of the present work is to compare different erosion models in Francis runner blades. For the study of the same, a single passage domain for the Francis runner blade was created using ANSYS Turbogrid. Further meshing was done on the same. To conduct the numerical study, OpenFOAM was used because it gives more freedom and optimization to the solver, where a user can compile according to specific needs. Comparisons for the Francis runner blades were made based on four different erosion models. The results obtained are very significant, considering that an analysis was done on the erosion patterns on the runner blade using various erosion models. On most occasions, the largest erosion happens to be usually at the pressure side, close to the trailing edge of the blade, thereby proving that this occurrence is a trend across all models. This pattern can then be attributed to greater relative velocities that exist in the runner's exit area. Noticeably, the models developed by Nandakumar and Arabnejad offer the unusual twist of pointing to an erosion pattern more pronounced on the suction side of the blade. In this departure from the body of literature, it shows that the models captured subtleties in erosion behavior that call for further research. Compared with the NandaKumar model and the Arabnejad Model, the DNV model and the Finnie model give more practical and closer predictions to previous study findings. The best match to the erosion pattern observed during the actual power plant site of Jhimruk Hydropower Plant is obtained with the result from the DNV model.

The results thus again underline the importance of the sediment-laden flow and its critical impact on the erosive attack of impellers and the use of numerical simulations for predicting patterns of erosion and designing better-performing turbines for sediment-rich conditions. Improved models and validation of such predictions over a larger operational condition and turbine geometry range furnish a need of further research.

Acknowledgements

The study was performed jointly at the Turbine Testing Lab of Kathmandu University and State Key Laboratory of Water Resources and Management, Wuhan University. The project was funded by Visiting Scholar's Fund provided by WRHES, China (Project No. 2021SDG01) and Hydro Himalaya project, Nepal.

References

1. Zoeb H, MohdZulkifly A, Zainal A. Basic fluid mechanics and hydraulic machines. BS Publications; 2008. 234 p.
2. Noon AA, Kim MH. Sediment and Cavitation Erosion in Francis Turbines—Review of Latest Experimental and Numerical Techniques. Vol. 14, Energies. MDPI AG; 2021.
3. Sangal S, Singhal MK, Saini RP. Hydro-abrasive erosion in hydro turbines: a review. Vol. 15, International Journal of Green Energy. Taylor and Francis Inc.; 2018. p. 232–53.
4. Qian Z, Zhao Z, Guo Z, Thapa BS, Thapa B. Erosion Wear on Runner of Francis Turbine in Jhimruk Hydroelectric Center. Journal of Fluids Engineering, Transactions of the ASME. 2020 Sep 1;142(9).
5. Rakibuzzaman M, Kim HH, Kim K, Suh SH, Kim KY. Numerical study of sediment erosion analysis in Francis turbine. Sustainability (Switzerland). 2019 Mar 1;11(5).
6. Truscott GF. Wear-Elsevier Sequoii A LITERATURE SURVEY ON ABRASIVE WEAR IN HYDRAULIC MACHINERY*.
7. Thapa B. Sand Erosion in Hydraulic Machinery. Ph.D. Thesis, Norwegian University of Science and Technology, 2004.
8. Bajracharya TR, Acharya B, Joshi CB, Saini RP, Dahlhaug OG. Sand erosion of Pelton turbine nozzles and buckets: A case study of Chilime Hydropower Plant. Wear. 2008 Feb 4;264(3–4):177–84.
9. Thapa BS, Thapa B, Dahlhaug OG. Empirical modelling of sediment erosion in Francis turbines. Energy. 2012;41(1):386–91.
10. Thapa BS, Dahlhaug OG, Thapa B. Sediment erosion in hydro turbines and its effect on the flow around guide vanes of Francis turbine. Vol. 49, Renewable and Sustainable Energy Reviews. Elsevier Ltd; 2015. p. 1100–13.
11. Chitrakar S, Neopane HP, Dahlhaug OG. A Review on Sediment Erosion Challenges in Hydraulic Turbines. In: Sedimentation Engineering. InTech; 2018. rvard

12. Neopane HP. Sediment Erosion in Hydro Turbines. PhD Thesis, Norwegian University of Science and Technology, 2010.
13. Lenarcic M, Eichhorn M, Schoder SJ, Bauer C. Numerical investigation of a high head Francis turbine under steady operating conditions using foam-extend. In: Journal of Physics: Conference Series. Institute of Physics Publishing; 2015.
14. Kang MW, Park N, Suh SH. Numerical Study on Sediment Erosion of Francis Turbine with Different Operating Conditions and Sediment Inflow Rates. In: Procedia Engineering. Elsevier Ltd; 2016. p. 457–64.
15. Zhao Z, Qian Z, Guo Z, Dong J. Erosion Wear on the Runner Shroud of a Francis Turbine. Journal of Energy Engineering. 2021 Dec;147(6).
16. Lopez A, Stickland M, Dempster W. Modeling erosion in a centrifugal pump in an Eulerian-Lagrangian frame using OpenFOAM®. Open Engineering. 2015 Jan 1;5(1):274–9.
17. Finnie I. Wear-Elsevier Sequoia S.A., Lausanne-Printed in the Netherlands SOME OBSERVATIONS ON THE EROSION OF DUCTILE METALS.
18. Peng W, Cao X. Numerical prediction of erosion distributions and solid particle trajectories in elbows for gas-solid flow. J Nat Gas Sci Eng. 2016 Mar 1;30:455–70.
19. López A, Stickland MT, Dempster WM. Comparative study of different erosion models in an Eulerian-Lagrangian frame using Open Source software.
20. Parsi M, Kara M, Sharma P, Gl D, Mclaury BS, Shirazi SA. OTC-27233-MS Comparative Study of Different Erosion Model Predictions for Single-Phase and Multiphase Flow Conditions. 2016.

IOP Conf. Series: Materials Science and Engineering **1314** (2024) 012010 doi:10.1088/1757-899X/1314/1/012010

Backtesting of value at risk estimation methods for energy commodity trading: evaluating performance and identifying the optimal approach

N Silwal[1,*] and S R Giri[2]

[1] Research Assistant, Department of Industrial Engineering, IOE, TU, Kathmandu, Nepal
[2] Head of Department, Department of Industrial Engineering, IOE, TU, Kathmandu, Nepal

E-mail: *nitesh.752421@thc.tu.edu.np

Abstract. This paper focuses on the application of Value at Risk (VaR) estimation techniques in the context of energy commodity trading, with a specific emphasis on the NYMEX (NG) and Title Transfer Facility (TTF) natural gas index. The primary point of this study is to break down and assess the VaR assessment strategies and recognize the most reasonable methodology. Historical price data for these commodities were collected, and forward prices were generated using a Long Short-Term Memory (LSTM) forecasting model. A portfolio comprising Future and Forward contracts for both NG and TTF index was constructed and three VaR estimation approaches: Variance-Covariance, Historical Simulation, and Monte Carlo Simulation were employed. Backtesting was then performed using the metric Quadratic Probability Score (QPS) to determine the optimal approach. Sensitivity analysis was also conducted to evaluate the variation in the estimated VaR values. The VaR values were determined using all three methods at different confidence interval and their backtesting was performed. The study found that the Monte Carlo Simulation approach outperforms other methods for estimating VaR in risk-averse trading scenarios with lowest QPS. It also underscores the need to consider individual contract risks within a portfolio, suggesting that diversification alone might not suffice in mitigating those risks.

1. Introduction

Financial disasters in the early 1990s have brought a lot of revolutions in the field of physical and financial markets, one being the introduction of Value at Risk (VaR) metric. This metric primarily uses the statistical approaches in the estimation of risk factor. It measures the maximum potential loss that a stock or a portfolio traded may face in the market. Overall, VaR is a forward centered risk calculation concept which basically makes the use of forward market prices and evaluates the historical patterns to quantify the possible future risks. Once the VaR quantification is done, it's precision is evaluated using backtesting, which involves the comparative analysis of actual losses and the predicted losses to validate the efficacy of the VaR model.

This paper incorporates a larger dataset comprising 843 days for the purpose of simulation. By increasing the number of data points for simulation, the enhanced approach allows for a more comprehensive analysis of historical price movements and better captures the underlying volatility and risk patterns in the market. This paper extends the use of the QPS metric to include two other approaches: Historical Simulation and Monte Carlo Simulation.

There are three popular approaches that use VaR metric for risk measurement: Variance-Covariance, Historical Simulation, and Monte Carlo Simulation. All these methods are best on their way of

IOP Conf. Series: Materials Science and Engineering **1314** (2024) 012010 doi:10.1088/1757-899X/1314/1/012010

calculation, and vary based on the nature of portfolio, available price data, and types of investment.

1.1. Variance-Covariance Approach:

This method makes the use of covariances (volatilities and correlations) of the forward price or the risk factors along with the portfolio variance to quantify the risk at given level of acceptance or the confidence [1, 2].

1.2. Historical Simulation Approach:

Historical simulation is a non-parametric approach of VaR estimation, which assumes that historical returns patterns are indictive of the patterns of the forward returns [3].

1.3. Monte Carlo Simulation Approach:

Monte Carlo VaR is based on the concept to repeatedly run a large number of simulations of random variables in order to cover a wide range of possible cases of determination of future returns [4].

Various research and studies have been conducted on determining the appropriate method for risk assessment of energy commodities. Debates regarding the best VaR estimation methods for commodity trading may be among the field of study's divergent and controversial hypotheses. Due to different assumptions and data requirements, various approaches, such as the Variance-Covariance, Historical Simulation, and Monte Carlo Simulation methods, may produce different outcomes. To generalize the outcomes of these approaches and compare them, we performed backtesting by making the use of Quadratic Probability Score (QPS) as discussed by [5]. In addition, our study also enhances the application of QPS to back test the optimal approach of VaR estimation by employing it for all three calculation approaches, by enhancing the study conducted by Puspa Liza Ghazali on [6] which employs the application of the QPS for variance-covariance approach only.

The primary point of this seminar study is to break down and assess the VaR assessment strategies for natural gas exchanging and recognize the most reasonable methodology. By examining real price data, employing advanced models, and conducting backtesting, this study aims to assess the accuracy and effectiveness of different VaR estimation techniques, as well as investigate the impact of VaR estimation on risk management strategies and portfolio performance in the natural gas market.

2. Materials and methodology

2.1. Data collection

The historical data from Yahoo Finance, using a Python script, had gathered pricing information for two natural gas indexes, NG=F (NYMEX, American Natural Gas Index), and TTF=F (Dutch Natural Gas Index). The data had comprised 843 trade days between January 1, 2020, and May 4, 2023.

With the intention of achieving greater diversification, a portfolio was created by incorporating indexes that encompassed both a low-volatility (NG=F) and a high-volatility (TTF=F) natural gas stock. The diversified portfolio had been utilized to efficiently calculate VaR. Four buy contracts, consisting of Future and Forward contracts for each index, had been included in the portfolio.

2.2. Data preparation

The monthly prices were obtained by averaging the daily prices, ensuring consistency in the VaR computation. This process allowed for the capture of broad pricing trends and patterns within a more manageable dataset. To obtain the forward prices required for VaR computation, the Long Short-Term Memory (LSTM) model, as described by [7], was employed for forecasting. The focus of the forecasting was on projecting the forward prices for the months of June, July, and August (a span of 3 months), with the forecasting period ranging from April 10, 2023, to May 4, 2023.

To prepare the data for forecasting, a 60-day lookback period was utilized as a time stamp. The dataset, consisting of a total of 843 days, was split into a 20% testing set and an 80% training set. The forecasting process of the model employed the Adam optimizer and ReLU activation function, while its performance was evaluated using the Mean Squared Error (MSE) measure. The model was trained over

200 iterations with a batch size of 32 for each epoch. These parameters were employed by the LSTM model in its attempt to accurately forecast the forward prices for the designated three-month period.

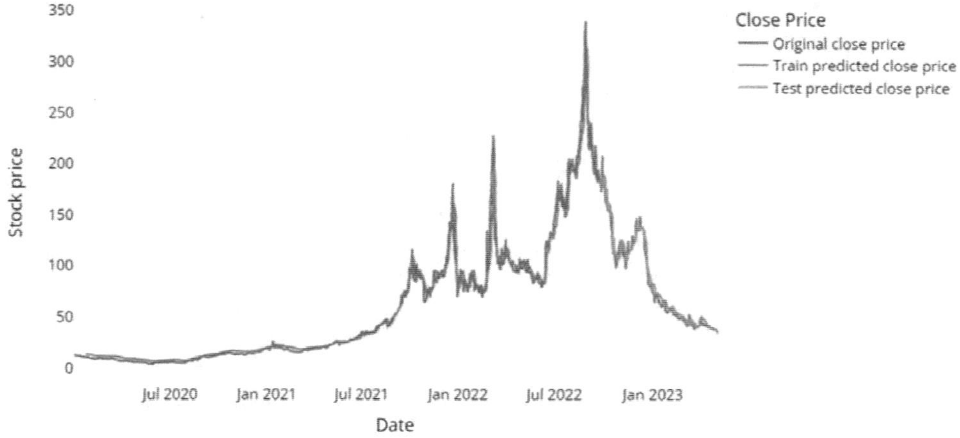

Figure 1. Original vs predicted forward price of natural gas index.

```
Model: "sequential"
_________________________________________________________________
 Layer (type)                Output Shape              Param #
=================================================================
 lstm (LSTM)                 (None, 10)                480

 dropout (Dropout)           (None, 10)                0

 dense (Dense)               (None, 1)                 11

=================================================================
Total params: 491
Trainable params: 491
Non-trainable params: 0
_________________________________________________________________
```

Figure 2. LSTM model summary.

2.3. VaR calculation

The calculation process started by assessing the volatility and correlation of the two price curves that represented the NG=F and TTF=F indexes. This analysis aimed to ascertain the extent of price fluctuations and the interrelationship between the indexes. Subsequently, the Mark-to-Market (MTM) value of the portfolio was determined, enabling an evaluation of its current market worth and financial standing.

To calculate the volatility and correlation matrix, assumption was made by the researcher that P_1, P_2,.....P_m represent the prices of the commodities, and V_1 , V_2,.....V_m represent their respective positions at specific terms of the contract from 1, 2,...,m and 'α' denotes the level of significance.

Historical price matrix
A matrix of historical prices for t days and m terms had been created, with the order of the matrix being m×t.

$$Prices = \begin{pmatrix} P_1 & \cdots & P_m \\ \vdots & \ddots & \vdots \\ P_t & \cdots & P_{m(t)} \end{pmatrix} \quad (1)$$

Rate of return matrix
The rate of return could be calculated by the logarithmic rate of return method.

$$r_t = ln\left(\frac{P_t}{P_{t-1}}\right) \quad (2)$$

So, a rate of return matrix could be represented as:

$$r_t = \begin{pmatrix} r_1 & \cdots & r_m \\ r_2 & & r_{2m} \\ \vdots & & \vdots \\ \vdots & \ddots & \vdots \\ r_{t-2} & & r_{m(t-2)} \\ r_{t-1} & \cdots & r_{m(t-1)} \end{pmatrix} \quad (3)$$

Volatility
Volatility is the measure of price movements or fluctuations on the market. It is the standard deviation of the returns.

$$volatility = (V_i, \quad V_{2i}, \quad \cdots \quad, V_{(m-1)i} \quad, V_{mi})_{i=1\ to\ (t-1)}$$

$$v = \left(\sqrt{\frac{(r_i - \bar{r}_i)^2}{t-1}}, \sqrt{\frac{(r_{2i} - \bar{r}_{2i})^2}{t-1}}, \quad \cdots \quad, \sqrt{\frac{(r_{(m-1)i} - \bar{r}_{(m-1)i})^2}{t-1}}, \sqrt{\frac{(r_{mi} - \bar{r}_{mi})^2}{t-1}}\right)_{i=1\ to\ (t-1)}$$

$$v = \begin{pmatrix} v_{1i} & 0 & 0 & \cdots & 0 \\ 0 & v_{2i} & 0 & \cdots & 0 \\ 0 & 0 & v_{3i} & \cdots & 0 \\ \vdots & \vdots & \vdots & \ddots & \vdots \\ 0 & 0 & 0 & \cdots & v_{mi} \end{pmatrix}$$

$$v = \begin{pmatrix} \sqrt{\frac{(r_i - \bar{r}_i)^2}{t-1}} & 0 & 0 & & 0 \\ 0 & \sqrt{\frac{(r_{2i} - \bar{r}_{2i})^2}{t-1}} & 0 & & 0 \\ 0 & 0 & \sqrt{\frac{(r_{3i} - \bar{r}_{3i})^2}{t-1}} & \cdots & 0 \\ \vdots & \vdots & \vdots & & 0 \\ 0 & 0 & 0 & & \sqrt{\frac{(r_{mi} - \bar{r}_{mi})^2}{t-1}} \end{pmatrix} \quad (4)$$

Correlation matrix
Correlation is the measure of degree of closeness or dependency between market curves. By using the rate of returns from the price curves from the equation, a correlation matrix could be generated.

$$c = \begin{pmatrix} 1 & c_{12} & c_{13} & \cdots & c_{1m} \\ c_{21} & 1 & c_{23} & \cdots & c_{2m} \\ c_{31} & c_{32} & 1 & \cdots & c_{3m} \\ \vdots & \vdots & \vdots & \ddots & \vdots \\ c_{(t-1)1} & c_{(t-1)2} & c_{(t-1)3} & \cdots & 1 \end{pmatrix} \tag{5}$$

The foundation for subsequent risk estimate and analysis had been laid by these first actions. Next, the VaR values for the portfolio were determined using three distinct methods: Variance-Covariance, Historical Simulation, and Monte Carlo Simulation. VaR estimates had been made for a 1-day holding time with a range of confidence levels, including 90%, 95%, and 99%.

2.3.1. Variance-covariance approach
The Variance-Covariance approach is a commonly used method for estimating Value at Risk (VaR). It relies on the assumption that asset returns follow a normal distribution. This approach calculates VaR by considering the variance of the asset returns and the covariance between the asset returns and the portfolio.

Firstly, the Variance-Covariance (VCV) matrix was calculated by multiplying the volatility matrix to the product of the correlation matrix and the volatility matrix to quantify the relationships and interactions between the assets in the portfolio. After obtaining the VCV matrix, the Mark-to-Market (MTM) matrix of the portfolio was computed by summing the MTM values of each asset sharing a common market or risk factor. This step provided an overview of the overall portfolio value. Using the VCV matrix and the assigned weights for each asset, the portfolio's volatility was determined. This measure helped assess the potential fluctuations and risks associated with the portfolio.

To estimate the Value at Risk (VaR), the portfolio volatility was multiplied by the square root of the holding period and the z-score value corresponding to the desired level of significance. This calculation provided an estimate of the potential loss the portfolio could experience within the specified holding period and confidence level.

Volatility-Correlation-Volatility (VCV) matrix:

$$VCV = \begin{pmatrix} v_i & 0 & 0 & \cdots & 0 \\ 0 & v_{2i} & 0 & \cdots & 0 \\ 0 & 0 & v_{3i} & \cdots & 0 \\ \vdots & \vdots & \vdots & \ddots & \vdots \\ 0 & 0 & 0 & \cdots & v_{mi} \end{pmatrix} \times \begin{pmatrix} 1 & c_{12} & c_{13} & \cdots & c_{1m} \\ c_{21} & 1 & c_{23} & \cdots & c_{2m} \\ c_{31} & c_{32} & 1 & \cdots & c_{3m} \\ \vdots & \vdots & \vdots & \ddots & \vdots \\ c_{(t-1)1} & c_{(t-1)2} & c_{(t-1)3} & \cdots & 1 \end{pmatrix} \times \begin{pmatrix} v_i & 0 & 0 & \cdots & 0 \\ 0 & v_{2i} & 0 & \cdots & 0 \\ 0 & 0 & v_{3i} & \cdots & 0 \\ \vdots & \vdots & \vdots & \ddots & \vdots \\ 0 & 0 & 0 & \cdots & v_{mi} \end{pmatrix} \tag{6}$$

MTM matrix

$$MTM\ Matrix = \begin{pmatrix} \sum_{i=1}^{m} (MTM)_1 \\ \sum_{i=1}^{m} (MTM)_2 \\ \sum_{i=1}^{m} (MTM)_3 \\ \vdots \\ \vdots \\ \sum_{i=1}^{m} (MTM)_{t-1} \end{pmatrix} \tag{7}$$

Portfolio Volatility

$$Portfolio\ Volatility = sqrt[\begin{pmatrix} \sum_{i=1}^{m}(MTM)_1 \\ \sum_{i=1}^{m}(MTM)_2 \\ \sum_{i=1}^{m}(MTM)_3 \\ \vdots \\ \vdots \\ \sum_{i=1}^{m}(MTM)_{t-1} \end{pmatrix}^T \times \left\{ VCV\ Matrix \times \begin{pmatrix} \sum_{i=1}^{m}(MTM)_1 \\ \sum_{i=1}^{m}(MTM)_2 \\ \sum_{i=1}^{m}(MTM)_3 \\ \vdots \\ \vdots \\ \sum_{i=1}^{m}(MTM)_{t-1} \end{pmatrix} \right\}] \quad (8)$$

Value at risk

$$VaR = Portfolio\ Volatility \times sqrt(holding\ days) \times (z\_score)_\alpha \quad (9)$$

2.3.2. Historical Simulation Approach
Historical Simulation is a risk management technique used to estimate the Value at Risk (VaR) of a portfolio or investment. It assumes that future returns and market conditions will be like historical data.

After the rate of return matrix had been calculated, the Historical Simulation approach was employed to estimate the Value at Risk (VaR) of the portfolio. The initial step involved determining the price weight matrix, which depicted the prices of the commodity on the specific date for which the VaR calculation was performed, commonly referred to as the 'as of date'. The matrix acquired possessed an order of 1*m. To simplify the calculation process, the transpose of this matrix was taken, resulting in an order of m*1. This transformation was conducted to facilitate subsequent calculations within the VaR estimation process.

$$w_p = (P_1 \quad \cdots \quad P_m)^T = \begin{pmatrix} P_1 \\ \vdots \\ P_m \end{pmatrix} \quad (10)$$

Following that, the curve value delta was determined by simulating the rate of return values for 'm' terms across 't' number of simulations. This step entailed calculating the Price Simulation, which was obtained by multiplying the exponential value of the rate of return for each term with the price weight matrix.

$$\delta P = w_p \times \exp(r_{t+1})$$

$$\delta P = \begin{pmatrix} P_1 \\ \vdots \\ P_m \end{pmatrix} \begin{pmatrix} e^{r_2} & \cdots & e^{r_m} \\ \vdots & \ddots & \vdots \\ e^{r_{t+1}} & \cdots & e^{r_{m(t)}} \end{pmatrix} \quad (11)$$

To evaluate the impact at the deal or contract level, the market value delta and contract value delta were calculated by multiplying the curve value delta of the market and contract price at a specific term with the position of the deal at the same term.

$$\delta M = (\delta P)_{market} \times \begin{pmatrix} V_1 \\ \vdots \\ V_m \end{pmatrix} \quad (12)$$

$$\delta C = (\delta P)_{contract} \times \begin{pmatrix} V_1 \\ \vdots \\ V_m \end{pmatrix} \tag{13}$$

By utilizing the market value delta and contract value delta, the Mark-to-Market (MTM) value matrix was calculated. This matrix represented the sum of the market value delta and contract value delta. The MTM value matrix aided in assessing the overall value and performance of the portfolio.

$$MTM = \delta M - \delta C$$

$$MTM\ matrix = \left\{ (\delta P)_{market} \times \begin{pmatrix} V_1 \\ \vdots \\ V_m \end{pmatrix} \right\} - \left\{ (\delta P)_{contract} \times \begin{pmatrix} V_1 \\ \vdots \\ V_m \end{pmatrix} \right\} \tag{14}$$

Lastly, to ascertain the VaR value, the MTM values for each simulation, spanning from 1 to t, were aggregated and sorted in ascending order. The position of the VaR value was determined by multiplying the desired level of significance with the number of look-back periods. This calculation provided a quantifiable measure of the potential loss at the specified confidence level.

Once, the MTM values are sorted in a column matrix, the position of VaR value was determined using the number of data points and the acceptance level.

$$Sorted\ sum\ of\ MTM = \begin{pmatrix} \sum_{i=1}^{m} (MTM)_1 \\ \sum_{i=1}^{m} (MTM)_2 \\ \vdots \\ \sum_{i=1}^{m} (MTM)_{t-1} \\ \sum_{i=1}^{m} (MTM)_t \end{pmatrix} \tag{15}$$

$$VaR\ position = level\ of\ significance(\alpha) \times t \tag{16}$$

2.3.3. Monte Carlo simulation approach

In addition to the Variance-Covariance and Historical Simulation techniques, Monte Carlo Simulation was utilized to calculate the portfolio's Value at Risk (VaR). The Monte Carlo Simulation method is effective in evaluating the potential outcomes of the portfolio through the generation of numerous random scenarios.

To begin, Cholesky Decomposition was performed by decomposing the positive definite correlation matrix into a lower triangular matrix. This decomposition allowed the transformation of independent random numbers into correlated random numbers, which is essential for accurately simulating the portfolio's performance.

$$L = \begin{pmatrix} \sqrt{c_{11}} & 0 & 0 & \cdots & 0 \\ \frac{c_{21}}{L_{11}} & \sqrt{c_{2i} - L_{21}{}^2} & 0 & \cdots & 0 \\ \frac{c_{31}}{L_{11}} & \frac{c_{31} - L_{31} * L_{21}}{L_{22}} & \sqrt{c_{3i} - L_{31}{}^2 + L_{32}{}^2} & \cdots & 0 \\ \vdots & \vdots & \vdots & \ddots & \vdots \\ \frac{c_{(t-1)1}}{L_{11}} & \frac{c_{2(t-1)} - L_{(t-1)1} * L_{21}}{L_{22}} & \frac{c_{3(t-1)} - L_{31} * L_{(t-1)1} - L_{32} * L_{(t-1)2}}{L_{33}} & \cdots & \sqrt{c_{m(t-1)} - L_{(t-1)1}{}^2} \end{pmatrix} \tag{17}$$

Following the Cholesky Decomposition, a matrix of normalized random numbers was generated. This matrix represented a set of random values with a standard normal distribution. By multiplying the Cholesky Decomposition matrix with the normalized random number matrix, "R", a correlated random number matrix "B" had been created. This step ensured that the generated random numbers retain the desired correlation structure as defined by the Cholesky Decomposition.

$$R = \begin{pmatrix} n_{11} & n_{21} & n_{13} & \cdots & n_{1t} \\ n_{21} & n_{22} & n_{23} & \cdots & n_{2t} \\ n_{31} & n_{23} & n_{33} & \cdots & n_{3t} \\ \vdots & \vdots & \vdots & \ddots & \vdots \\ n_{m1} & n_{m2} & n_{m3} & \cdots & n_{mt} \end{pmatrix} \tag{18}$$

$$N = \begin{pmatrix} NORMS.INV(n_{11}) & NORMS.INV(n_{12}) & NORMS.INV(n_{13}) & \cdots & NORMS.INV(n_{1t}) \\ NORMS.INV(n_{21}) & NORMS.INV(n_{22}) & NORMS.INV(n_{23}) & \cdots & NORMS.INV(n_{2t}) \\ NORMS.INV(n_{31}) & NORMS.INV(n_{23}) & NORMS.INV(n_{33}) & \cdots & NORMS.INV(n_{3t}) \\ \vdots & \vdots & \vdots & \ddots & \vdots \\ NORMS.INV(n_{m1}) & NORMS.INV(n_{m2}) & NORMS.INV(n_{m3}) & \cdots & NORMS.INV(n_{mt}) \end{pmatrix} \tag{19}$$

$$B = L \times N \tag{20}$$

Utilizing the weighted price matrix, volatility matrix, and correlated random number matrix, a price simulation was performed for 1000 data points. This simulation process led to the generation of the Simulated Prices Matrix, which provided a collection of simulated price values based on the given inputs and the random scenarios generated.

$$S_{ij} = P_i \times exp\,(Drift - \frac{{v_i}^2}{2} + v_i \times b_{ij})_{i=1\ to\ m, j=1\ to\ t}$$

$$S_{ij} = \begin{pmatrix} s_{11} & s_{21} & s_{13} & \cdots & s_{1t} \\ s_{21} & s_{22} & s_{23} & \cdots & s_{2t} \\ s_{31} & s_{23} & s_{33} & \cdots & s_{3t} \\ \vdots & \vdots & \vdots & \ddots & \vdots \\ s_{m1} & s_{m2} & s_{m3} & \cdots & s_{mt} \end{pmatrix} \tag{21}$$

From the simulated prices matrix, curve value delta matrix was determined by subtracting it from the weighted price matrix.

$$\delta P = S_{ij} - w_p$$

$$\delta P = \begin{pmatrix} P_1 \\ P_2 \\ P_3 \\ \vdots \\ P_m \end{pmatrix} - \begin{pmatrix} s_{11} & s_{21} & s_{13} & \cdots & s_{1t} \\ s_{21} & s_{22} & s_{23} & \cdots & s_{2t} \\ s_{31} & s_{23} & s_{33} & \cdots & s_{3t} \\ \vdots & \vdots & \vdots & \ddots & \vdots \\ s_{m1} & s_{m2} & s_{m3} & \cdots & s_{mt} \end{pmatrix} \tag{22}$$

After obtaining the curve value delta matrix, the subsequent steps follow a similar process as the historical simulation approach as defined from equation (12) to equation (16). This includes determining the market value delta matrix, calculating the Mark-to-Market (MTM) value matrix, and estimating the Value at Risk (VaR) using the sorted MTM values. The steps involved in these calculations remain consistent with the methodology employed in the historical simulation approach.

2.4. Backtesting the VaR models

The method of backtesting VaR models is used to evaluate the precision and dependability of VaR estimations. It entails contrasting the projected VaR values with the actual results shown in historical data or during a predetermined time frame. Backtesting is done to assess how well VaR models perform in terms of calculating and forecasting the risk associated with a portfolio or financial instrument.

In this study, the backtesting of the previously described VaR models was conducted using the Quadratic Probability Score (QPS) metric, following the forecast evaluation approaches proposed by [5]. The proposed backtesting methodology utilizes a loss function that consists of one output and four essential components. The first component comprises a set of n observations of loss or gain for each period, which are linked to the corresponding VaR projections. The second component is the value of the loss function, which is influenced by the VaR forecasts over the relevant time. The specific formulation of the loss function, as defined by [5], is provided below:

$$C_t = \begin{cases} \frac{L_t - VaR_t}{VaR_t}, & L_t > VaR_t \\ 0, & L_t \leq VaR_t \end{cases} \tag{23}$$

where, L_t is the loss or profit value.

The third component of the loss function is the benchmark value, which represents the expected value of the best model. The fourth component is the value of the function, which serves as the input for both the reference value and the loss function. The Quadratic Probability Score (QPS) is then calculated using these components.

$$QPS = \frac{2}{n}\sum_{t=1}^{n}(C_t - p)^2 \tag{24}$$

where, p = 1 – α and n denotes number of sets of data under analysis. According to the researcher, the QPS values range from [0, 2], with lower values indicating improved model performance.

2.5. What if analysis

After backtesting and selection of the best model for VaR calculation, a what-if study was conducted to validate the predicted outcomes and analyze the impact on the VaR value of the model. In this study, a hypothetical scenario was created where the price of the commodity in the portfolio experienced a 20% change in either an upward or downward direction. The changes in the VaR value corresponding to these simulations were assessed, providing valuable insights into the sensitivity and risk exposure of the portfolio under various pricing scenarios which helped to understand the potential risks and implications of significant price movements on the portfolio's VaR.

3. Results and discussion

3.1. Forward price forecasting with LSTM model

The strong alignment between the training and testing losses indicated excellent performance of the model. The convergence between these losses suggested that the model was effectively trained and capable of making accurate future price predictions. The minimal losses obtained demonstrated the model's ability to accurately identify patterns and trends in the data, highlighting its reliability and robustness in price predictions.

Below is a representation of the different loss metrics obtained during the training and testing phases of the model and a plot of training and testing loss of the used LSTM model.

Table 1. Loss and accuracy metrices for train and test data

Metrics	Train	Test
Mean Absolute Error	0.143447	0.355608
Mean Squared Error (MSE)	0.002833	0.003120
R^2 Score	0.971033	0.952043

3.2. VaR Calculation

Indeed, constructing a well-defined portfolio with a diversified mix of assets and instruments is a crucial initial step in VaR calculation. Once the portfolio composition is established, the next stage involves doi:10.1088/1757-899X/1314/1/012010

evaluating the current market prices of each asset or instrument held within the portfolio. This process, referred to as Mark-to-Market (MTM), aims to determine the fair value of the portfolio based on the prevailing market values. The MTM procedure is essential as it provides an accurate depiction of the portfolio's current value. The MTM value serves as a vital input for subsequent VaR computations, allowing for a comprehensive assessment of the potential risk and exposure associated with the portfolio. By considering the current market prices through MTM, the VaR calculations can reflect the real-time value and dynamics of the portfolio accurately. MTM calculation result for the defined portfolio is tabulated below:

Table 2. Mark to market value of the defined portfolio

Contract	Term Start	Term End	Position	Forward Price	Contract Price	Forward Value	Contract Value	Mark to Market
NG=F Forward	6/1/2023	6/30/2023	5000	2.53	2.7	12,631.73	13,500	-868.27
	7/1/2023	7/31/2023	5000	3.16	2.7	15,813.13	13,500	2,313.13
	8/1/2023	8/31/2023	5000	3.42	2.7	17,095.64	13,500	3,595.64
NG=F Future	6/1/2023	6/30/2023	4500	2.53	2.7	11,368.56	12,150	-781.44
	7/1/2023	7/31/2023	4650	3.16	2.7	14,676.35	12,555	2,121.35
	8/1/2023	8/31/2023	4650	3.46	2.7	16,065.93	12,555	3,510.93
TTF=F Forward	6/1/2023	6/30/2023	5000	47.47	50	237,369.24	250,000	-12,630.76
	7/1/2023	7/31/2023	5000	80.27	50	401,368.98	250,000	151,368.9
	8/1/2023	8/31/2023	5000	94.53	50	472,649.99	250,000	222,649.9
TTF=F Future	6/1/2023	6/30/2023	4500	47.47	50	213,632.31	225,000	-11,367.69
	7/1/2023	7/31/2023	4650	79.94	50	371,739.91	232,500	139,239.9
	8/1/2023	8/31/2023	4650	94.53	50	439,587.63	232,500	207,087.6
						2,223,999.4	**1,517,760**	**706,239.4**

Also, the VaR values calculated from three different approaches at different confidence level are attached in the table below.

Table 3. VaR values obtained from different approach at different confidence level in $.

Confidence Level	Variance Covariance	Historical	Monte Carlo
90%	91,528.075	1,490,824.92	1,616,088.736
95%	1,179,859.667	1,651,880.401	1,727,585.148
99%	1,666,104.863	398,203.976	1,901,076.327

The probability distribution for Monte Carlo Simulation at 95% confidence level is also attached below:

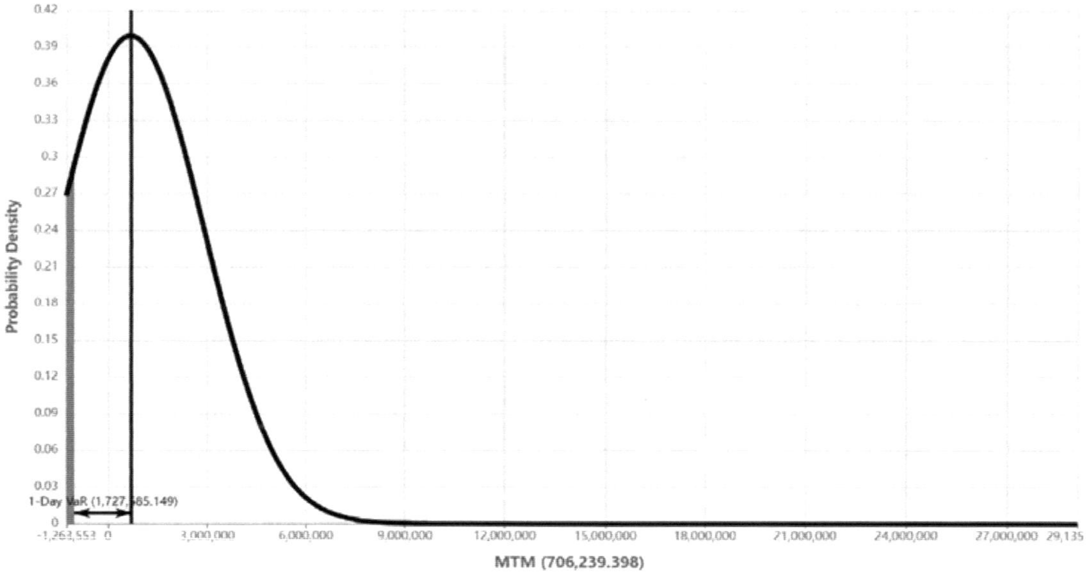

Figure 3. Probability distribution chart of Monte Carlo simulation at 95% confidence level

3.3. Expected shortfall

The average of all losses above the VaR threshold is known as Expected Shortfall or the Component VaR (CVaR). It measures the anticipated size of losses in the tail of distribution and offers information on the possible severity of extreme occurrences or unfavorable market circumstances. The expected shortfall obtained because of VaR estimation for each approach at different level of confidence are tabulated below:

Table 4. Expected shortfall values at different confidence level in $

Confidence Level	Variance Covariance	Historical	Monte Carlo
90%	1,253,410.371	1,640,926.2	1,750,224.182
95%	1,479,591.618	1,659,183.2	1,826,129.022
99%	1,908,797.071	0	1,945,942.84

3.4. Backtesting VaR models

Based on the backtesting model proposed in equations (23) and (24), the Loss function and the QPS value for each approach was calculated and hence the results are shown below:

Table 5. Loss function values of different models at different confidence level.

Confidence Level	Variance Covariance	Historical	Monte Carlo
90%	6.7161	0	0
95%	0	0	0
99%	0	0.7736	0

Table 6. Quadratic Probability Score of different models at different confidence level.

Confidence Level	Variance Covariance	Historical	Monte Carlo
90%	2.8189	**0.0752**	0.0817
95%	0.0752	0.0817	**0**
99%	0.0817	0.7736	**0**

3.5. Comparison of portfolio VaR with individual contract VaR

The VaR values of individual contracts of the portfolio were also calculated and compared with the portfolio VaR estimated by the outstanding approach i.e., Monte Carlo Simulation at 95% confidence level. The results are tabulated below:

Table 7. Individual VaR values of the portfolio at 95% confidence level in $.

Contracts	VaR Value ($)
NG=F Forward	13,277.58
NG=F Future	12,272.59
TTF=F Forward	898,575.30
TTF=F Future	831,748.90
Sum of individual VaR	**1,755,874.37**
Portfolio VaR	**1,727,585.148**

To represent this scenario mathematically, the individual VaR values of the contracts were assigned as VaR_1, VaR_2, VaR_3, and VaR_4.

The overall VaR of the portfolio, denoted as VaR_Portfolio, refers to the estimated potential loss for the entire portfolio at a specific confidence level, which is 95% in this case.

The equation introduced below:

$$VaR_1 + VaR_2 + VaR_3 + VaR_4 > VaR\_Portfolio \tag{25}$$

expresses the concept that the sum of the individual VaR values (the parts) should be greater than the VaR of the overall portfolio (the whole) to indicate that the combined risk or potential loss of the individual contracts is higher than the risk associated with the portfolio.

3.6. What If scenario analysis

When a scenario of both increment and decrement of the existing price of the natural gas was analyzed between -20% to +20%, the change in VaR values was also obtained by the same rate.

Table 8. What if analysis results

Price Change (%)	VaR Value ($)
-20%	1,382,068.12
-10%	1,554,826.63
-5%	1,641,205.89
0%	1,727,585.15
5%	1,813,964.41
10%	1,900,343.66
20%	2,073,102.18

When assessing various VaR calculation methods across three distinct confidence levels using the QPS metric, it was observed from that, apart from the Variance-Covariance method at a 90% confidence level, the QPS values for all other methods fell within the range of [0,2]. Among these, the lowest QPS values were associated with the Historical Simulation method at a 90% confidence level and the Monte Carlo Simulation method at both 95% (single tailed probability distribution curve in Figure 3) and 99% confidence levels. Consequently, it can be said that the Monte Carlo Simulation approach is the optimal choice for risk-averse trading and investment, as it consistently exhibits low QPS values within the preferred range.

Also, a notable observation from Table 7 is that the sum of individual Value at Risk (VaR) values exceeds the VaR of the overall portfolio. This suggests that the collective risk posed by the individual contracts is greater than the risk of the portfolio when considering them together. This finding raises questions about the effectiveness of portfolio diversification in mitigating risks associated with individual contracts.

The results of sensitivity analysis on Table 8 also establishes a direct relationship between changes in commodity prices and corresponding changes in VaR values. This implies that as the price of the commodity increases or decreases, the VaR values of the portfolio also increase or decrease accordingly. This highlights the sensitivity of VaR to commodity price fluctuations and emphasizes the need for monitoring and managing price risks within the portfolio.

4. Conclusions

From Table 5, it is evident that the Monte Carlo model achieves perfect loss function outcomes at the 90% and 95% confidence levels, with no losses recorded. Meanwhile, the Historical model also performs exceptionally well at the 90% confidence level, recording zero losses. However, at the 99% confidence level, the Monte Carlo model slightly edges out the competition with a minimal loss of 0.7736, indicating its robustness even at higher confidence levels.

Moving on to Table 6, which presents the Quadratic Probability Score (QPS), the Monte Carlo model continues to demonstrate superior performance across all confidence levels. At the 90% confidence level, the QPS for the Monte Carlo model is 0.0817, which is slightly higher than the Historical model but still indicates a very low risk score. Interestingly, at the 95% and 99% confidence levels, the Monte Carlo model achieves a perfect QPS score of 0, further highlighting its efficiency and accuracy in predicting risk at these confidence intervals.

Thus, the Monte Carlo model stands out as the most effective among the three models evaluated, demonstrating near-perfect outcomes in terms of both loss function values and Quadratic Probability Scores across all confidence levels. Its ability to minimize losses and achieve low risk scores makes it an ideal choice for applications requiring high reliability and precision in risk assessment.

Expanding the analysis beyond volatile natural gas commodities to include other commodities such as Renewables, Oils, Coal, and more would provide a comprehensive understanding of the risks associated with different energy products and their impact on portfolio performance. Incorporating a diverse set of commodities in future research would enhance the generalizability of the findings and provide insights into the specific risk characteristics of various commodities. This broader analysis would contribute to a more comprehensive perspective on portfolio risk management in the energy sector. Also, the future research can be made on the scope of improvement in volatility calculation methods like Exponential Weightage Moving Average (EWMA) and GARCH (1,1) methods.

Overall, these findings contribute to our understanding of risk management in energy portfolios and offer insights into the effectiveness of different approaches, the importance of diversification, and the influence of commodity price changes on portfolio risk.

Acknowledgements

The authors are very thankful to Department of Industrial Engineering, IOE, Thapathali Campus for providing help and support during this research.

IOP Conf. Series: Materials Science and Engineering **1314** (2024) 012010 doi:10.1088/1757-899X/1314/1/012010

REFERENCES

[1] Deutsch H P and Beinker M W 2009 *Derivatives and Internal Models* (London: Palgrave Macmillan) pp 430-61

[2] Walia J 2021 *The variance-covariance method for VaR calculation*

[3] Sharma M 2012 *The Historical Simulation Method for Value-at-Risk: A Research Based Evaluation of the Industry Favorite*

[4] Walia J 2022 *The Monte Carlo simulation method for VaR calculation*

[5] Blanco C and Ihle G 1998 How good is your VaR? Using backtesting to assess system performance *Financial Engineering News August* pp 1-2

[6] Ghazali P, Riaman R and Ulfatmi R 2020 Calculation of Value-at-Risk Variance-Covariance with the Approach of Simple Cash Portfolio, Factor Models and Cash Flow *Operations Research : International Conference Series* pp 22-3

[7] Elsworth S and Güttel S 2020 Time Series Forecasting Using LSTM Networks:A Symbolic Approach *ResearchGate* pp 5-6

Performance analysis of the blade for the v-shaped vertical axis wind turbine

Aastik Sharma[1,2], Lei Zhang[1], Jianjun Qu[1,4] and Sagar Panthi[3]

[1] School of Mechatronic Engineering, Harbin Institute of Technology, Harbin, China
[2] Department of Civil Engineering, Kathford International College of Engineering & Management, Lalitpur, Nepal
[3] Department of Mechanical Engineering, Beijing Institute of Technology of Technology, Beijing, China
[4] Author to whom correspondence should be addressed: qujianjun@hit.edu.cn

Abstract. A large-scale vertical axis wind turbine (VAWT) can lower the installation, operation, and maintenance cost for offshore areas as compared to the current offshore horizontal axis wind turbine (HAWT) designs. This is because of its simplicity of design and the location of heavy and complex mechanical parts near the water's surface. On that account, this article studies the aerodynamic performance capacity- the ability to efficiently capture and convert wind energy into rotational mechanical energy- of different blade models for novel V-shaped VAWT It has always been a challenging task to formulate a mathematical model for the complex aerodynamic behavior of the air around the VAWT blades. Nevertheless, previous studies have shown that prediction of the performance has been done more accurately using the CFD model compared to other models. Hence, a similar approach is used for the blade models in this research. The article studies the change in aerodynamic performance of V-VAWT with the change in the shape of its blade using 3D CFD simulation. A transient RANS simulation is carried out for different models of blades. For this, a hybrid mesh model is developed and simulations are conducted in ANSYS Fluent. The mesh independence, time-step independence, and rotation independence tests are completed to reduce errors in the simulation. The coefficient of moment (Cm) developed by the blade for five different blade models is studied. The aerodynamic performance of higher solidity constant chord blades is found to be higher than lower solidity constant chord blades. Also, the performance of tapered blades is found to be better than the constant chord blade, and tapered blades are suggested for the V-VAWT turbine models.

Keywords: VAWT, aerodynamics, blade design, CFD

1. Introduction

Conventional HAWT designs for offshore deployment have inherent difficulties with its ever-increasing size. The bulky gearbox and generator placed at the top of the turbine tower shift the center of gravity upward requiring a robust substructure for the stability of HAWT. The location of complex mechanisms at the top of the turbine also significantly increases operation and maintenance costs (1). The VAWT, on the other hand, could be a better alternative in offshore areas as heavy parts are located at the turbine's base, consequently requiring an economical substructure(2). Their omnidirectional ability doesn't require a pitching and yawing mechanism, reducing the complexity. The ease of accessibility and simple designs of VAWT aid in lowering operation and maintenance costs in offshore areas. Due to these factors, VAWT could harvest wind energy sustainably in the deep offshore regions where floating structures are required. Hence, the research on VAWT is growing to exploit the untapped high-quality wind sources in the offshore areas (3).

The main objective of the research on VAWT would be the reduction of the Levelized Cost of Energy (LCOE)(4). LCOE is the cost of generating electricity from a source, including all the expenses incurred during its lifetime. The LCOE of offshore wind projects is still high (0.45 USD/kWh) as compared to onshore wind projects (0.13 USD/kWh) (5).To reduce LCOE at offshore areas, the rotor design of VAWT could play a significant role. This is because the cost of the rotor itself is not as much when compared to the turbine's life cycle cost: it is 7% of turbine budget and the cost of turbine is only 28% of total life-cycle cost(6). But, an optimized rotor design can reduce the total budget of the wind project. The rotor design broadly includes aerodynamic and structural design. The article studies the aerodynamic capacity of different blade models for a novel V-shaped VAWT. The V-shaped VAWT is a towerless lift-driven VAWT model with the blades fixed at the base of the turbine. The blades are inclined at a certain degree to the vertical axis. The blade in this article uses NACA 0012 airfoil profile, as mostly symmetric airfoils of the NACA series have been used in the past(7). The higher-performing airfoils are also being studied(8) and could be used in future optimization of the blades.

There has been multiple research on the inclined blade for the VAWT(9–12). The comprehensive study of the large-scale VAWT at offshore areas conducted by Sandia National Laboratory (SNL) recently drew some meaningful results (4). The report extensively compares different configurations of Darrieus and V-shaped rotors. In the report, depending on chord length and number of blades, the power coefficient (Cp) for the V-shaped rotor is calculated to be in the range of 0.46-0.50. The lower Moment of Inertia (MOI) is obtained for the V-shaped designs than the Darrieus designs. Also, soft flutter modes are observed for the V-VAWT, while hard flutter modes are noticed for the Darrieus VAWT. This shows the potential of inclined blade application for the VAWT.

For a lift-driven VAWT, the forces developed on the blade are inconsistent because of changing relative velocity (W) and angle of attack (AOA) as the blade rotates around its axis. The changing W and AOA results in fluctuating lift and drag forces, which are responsible for producing a tangential force that helps to generate torque for the rotation of the blade. The tangential force is also unsteady depending on the azimuthal position. This is the reason why torque ripple is generated in the VAWT. The estimation of the forces is required to figure out the aerodynamic performance of the VAWT blade. This can be done by resolving the flow field around the blade.

The aerodynamic performance analysis of the VAWT has been performed using different engineering models such as the Streamtube model, cascade model, actuator cylinder model, vortex model, panel model, and CFD (13,14) The computational time and fidelity of the models vary from each other(15). The widely used Double Multiple Streamtube model is helpful for quick estimation of the performance with lower accuracy(11,16,17). On the other hand, CFD simulation requires significant computational resources but provides solutions of higher fidelity. Focusing on higher accuracy, CFD simulation is chosen to study the blade shape of VAWT. However, the meshing and simulation of a large-scale blade model would require significant time and computation resources. This is because, for a larger blade, a considerable number of cell elements are required to accurately capture the blade's geometry and resolve the boundary layers. The computation time will increase remarkably for the larger blade as the number of mesh increases. Regarding this fact, the blade models of V-VAWT were studied for the smaller size of the blade as accurate results can be obtained with fewer mesh elements than the large-scale blade. This would allow us to get CFD simulation results in a timely manner. Therefore, the blade models are developed for a 100 W rotor. The blade is inclined at a 35-degree inclination angle to the vertical axis (18).

Detailed CFD analysis for VAWT is often performed on 2D because of the requirement of a large number of grids and the consequent increase in computational cost. The Mesh independency technique was studied by Almohammadi et al. for 2D VAWT(19). The study mentioned several other mesh independency tests performed before. It concluded that a large number of meshes would be required to achieve a mesh-independent solution even for a 2D case. Another 2D simulation was performed by B. Hand et al. to study the dynamic stall on an airfoil at high Reynolds number flow, analyzing the effect of using different turbulence models, primarily targeted large-scale VAWT. The research concluded that at high TSR flow, the dynamic stall will be absent, and coarser grids are sufficient to calculate the performance of VAWT(20). The guidelines for selecting domain size and azimuthal increment in the case of 2D CFD are provided by Rezaeiha et al. (21). Bangga et al. has performed extensive research on

CFD simulation of three different VAWT using an engineering model and three additional CFD software i.e. FLOWer, TAU, and Ansys Fluent, which provides comparisons of results on that CFD software as well as comparison on DMST, vortex, and CFD approach (22).R. Howell et al.'s study of a small-scale VAWT, both numerically and experimentally, shows a reasonable agreement of 3D CFD results with experimental measurement(23). The model considered the effect of turbine solidity, tip vortices, and roughness using the RNG $k-\epsilon$ turbulence model. Zhang and et al. studied a three-bladed H-typed Darrieus wind turbine by performing a 3D simulation(24). Alaimo et al. did a 3D CFD analysis of four different orientations of 1.15 m 3-bladed VAWT using about 10 million cells which provides a basis for the selection of domain size and boundary condition applied in VAWT(25). These 2D and 3D CFD approaches provide guidelines for selecting solution parameters in this paper.

2. Methodology for CFD of VAWT

CFD is a complex numerical method basically trying to solve governing differential equations of how fluid behaves. For wind turbines, CFD numerically solves Naiver-Stokes equations for incompressible and isothermal fluid flow. Energy conservation doesn't play a role in the calculation. The original continuity and N-S equation in differential form for incompressible flow are given in equation (1) and (2)

$$\nabla.(\rho u) = 0 \tag{1}$$

Where ∇ is a del operator, ρ is density, and $u(x,t)$ is velocity. And, for x-direction,

$$\frac{\partial u}{\partial t} + u.\nabla u = -\frac{1}{\rho}\nabla \text{p} + \nu(u.\nabla)u + \text{g} \tag{2}$$

Where, $\nu = \mu/\rho$ is kinematic viscosity, $p(x,t)$ is pressure, and $g(x,t)$ is the body accelerations per unit mass. (e.g. gravity). The equation for the y and z-direction can be written similarly.

The above differential equations are discretized into sets of algebraic equations after including turbulence parameters. Reynolds average Navier-Stokes (RANS) equations was solved after modeling turbulence using $k-\omega$ SST model (22,25). The $k-\omega$ SST model solves Reynolds stress components in RANS using turbulent kinetic energy (ϵ) and specific dissipation rate (ω). The continuity and N-S equation including these two terms, were solved iteratively using pressure-based SIMPLE algorithm. The discretized algebraic equations from the RANS method represent a discrete spatial location. Hence, a CFD mesh model was developed as a part of preprocessing for the CFD simulation(26).

2.1 *Model development*

The model development started with a size estimation of the blade inclined at $35°$ for the $100\ W$ rotor. The required swept area(A) was taken to be $0.602\ m^2$ after the initial estimation from equation (3). The density of air(ρ) and design velocity V_d were taken as $1.225\ kg/m^3$ and $10\ m/s$, respectively, assuming a performance coefficient (Cp) of 0.3.

$$Cp = \frac{Power}{\frac{1}{2} \times \rho \times A \times V_d^3} \tag{3}$$

The moment coefficient (Cm) for VAWT was computed from equation (4), which provides the total torque generated that allows blade rotation.

$$Cm = \frac{Torque}{\frac{1}{2}\rho V_\infty^2 AR} \tag{4}$$

From equation (3) and (4), Cp can also be written as,

$$Cp = Cm * TSR \tag{5}$$

Where $TSR = \omega R/V_\infty$, is the Tip speed ratio of the turbine

For a V-shaped rotor with a base radius r_1, the swept area was calculated as:

$$A = (r_1 + R) \times H \tag{6}$$

The tip radius 'R' and vertical height 'H' of the turbine were calculated by applying the geometric relation for the inclined blade. The base radius r_1 was assumed to be 0.15 m, taking reference to the assumption from Soraghan (18).In doing so, the length of blade L was estimated to be 0.9m long using equation(6).

The blade shape was prepared in CATIA V5 software using a generative shape design after importing airfoil coordinates for NACA 0012.The meshing around the blade surface was completed in Pointwise meshing software. The meshing of the blade surface was done using a hybrid meshing technique, in which the boundary layers on the blade wall were grown using prism cells, and the far-field region was filled with tetrahedral cells. The structured meshing on the blade surface allowed to capture the airfoil's sharp leading and trailing edge surface. The unstructured meshing on the far-field region helped complete the meshing process quickly with fewer cell elements.

The CFD mesh model was broadly divided into the inner rotating domain, consisting of the blade, and the outer fixed domain, as shown in **Figure 1**. The shape and dimension of the inner and outer domains is presented in **Table 1**. The size and location of the inner and outer domains are specified according to Rezaeiha and et. al.(27).The location of the inner domain with respect to outer domain is also specified in **Table 1**. The inner and outer domains with the non-conformal interface were connected using the sliding mesh interface. For the inner domain, the mesh size around the blade walls was required to be significantly small. Therefore, the meshing process begins with the estimation of the first cell height needed to resolve the boundary layers on blade walls. Assuming a relative velocity of 60 m/s, the first cell height to satisfy $Y+=1$ was calculated to be 0.005 mm for a 100 mm chord length in www.pointwise.com/yplus/ . Here Y+ is the non-dimensional wall distance which is supposed to be less than 1 so the first cell height lies in the viscous sublayer. The blade's airfoil cross-section was divided into 72 nodes. To resolve the curvature of leading and trailing edges, nodes on leading and trailing edges were distributed with a growth rate of 1.2, maintaining a 0.3 mm distance at the edges. The blade was divided into 150,200, and 250 divisions along the span (0.9 m) to create different mesh model. Blade consisted higher-density nodes at both edges. The top surface of the blades on both sides was divided using an unstructured mesh. The boundary layers were grown on all blade surfaces using 30 inflation layers. The growth rate of 1.15 was applied for the first 15 layers, then 1.25 for another ten layers, and 1.3 for the remaining 5, respectively. The remaining space of the inner domain was populated with tetrahedral mesh. The transition between the hexahedral cells and tetrahedral cells was done using pyramid cells. The details of meshing in the inner domain is presented in **Figure 2**.

Four different mesh models were created for mesh independence test. Mesh A consisted of 150 divisions along the blade span with all other details kept as mentioned above. Mesh B was created just like Mesh A with 200 divisions along the blade. Mesh C was created by increasing the density of unstructured cells outside of inflation layers in Mesh B model. Mesh D was similar to Mesh A with 250 divisions along the blade span. The meshing of the outer fixed domain was done using the unstructured grid. The orthogonal quality of the overall mesh was above 0.001. The skewness of the mesh was below 0.8. Thus, maintaining a good overall quality of a mesh.

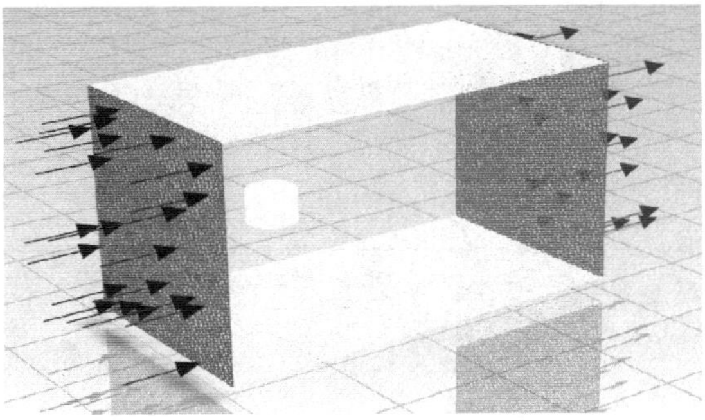

(a)

Figure 1. CFD model showing inner and outer domain.

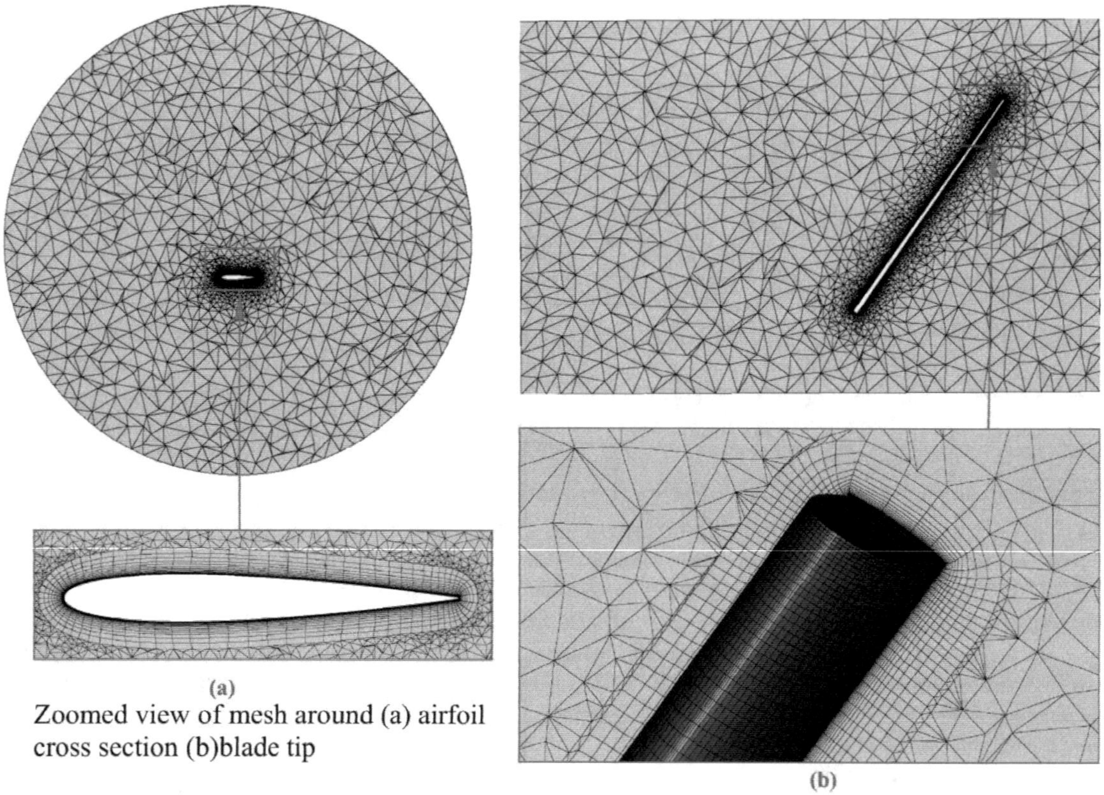

Figure 2. Meshing details of inner domain using top sectional and side sectional view including zoomed details of mesh around blade.

Table 1. Details of CFD domain.

Domain	Dimension
Inner rotating domain	Cylindrical: Radius= 1m, height= 1.3m (Mesh B:1,212,595 cells)
Outer domain	Cuboid: width= 8m, height 8m, length 16m (4m from the inlet and 12 m to the outlet from the center of rotation) (520,093 cells)

2.2 Solution procedure

The spatial and temporal discretization of the model was done using a second-order upwind scheme and second-order implicit scheme, as these schemes are unconditionally stable irrespective of mesh size and time step, respectively. The transport quantities on the boundary faces were calculated using the least square cell-based method. The convergence of the variables was improved using explicit under-relaxation factors. **Table 2** shows a detailed list of parameters selected for the CFD solution.

Table 2. CFD solution parameters.

Material	Standard properties of air, $\rho = 1.225\ kg/m^3, \mu = 1.7895 \times 10^{-5}\ kgm^{-1}s^{-1}$
Cell zone condition	The inner rotating domain is rotated at an angular velocity, $\omega = TSR * \frac{V_\infty}{R}$
Boundary Conditions	Inlet: Velocity inlet (10 m/s in x-direction) Turbulent intensity = 5% Turbulent viscosity ratio: 10 Outlet: Pressure outlet

	Turbulence parameter is the same as in Inlet Top, bottom, and side face of outer domain: Symmetry BC with zero-shear slip walls Blade walls: Zero-shear slip walls with no relative motion with adjacent cell zone. Between inner and outer domain: Sliding mesh Interface
Explicit under-relaxation factors	Pressure: 0.3, Density: 1, Body force: 1. Momentum: 0.5, Turbulent kinetic energy: 0.6, Specific dissipation rate 0.6; Turbulent viscosity: 0.6
Reference value	Area: $0.602\ m^2$, Radius: $0.66622m$, Velocity: $10\ m/s$
Residual parameter	Residual set to 10^{-3} for continuity equation and 10^{-5} for other residuals
Time step	The time step for the 1-degree rotation of the blade is calculated as: $$Time\ step\ (T) = \frac{\pi}{180 * \omega}$$

The inner domain was rotated clockwise, and the transient simulation was performed with at least 30 iterations for each time step. For the convergence criteria of the solution, apart from the residual parameter, the Coefficient of the moment (Cm) produced by the blade along the central z-axis was calculated from the simulation.

2.3 *Factors affecting CFD simulation*

The solution of CFD simulation could be inconsistent depending on several factors. The references from the previous research and careful selection of the parameters were carried out to ensure the accuracy of the solution. In addition, the resolution of the boundary layer, the density of the mesh, the time given to fully develop the flow, and the time step of the transient simulation were examined to obtain an accurate solution. The initial tests were performed on a 50-50 mm constant-chord blade model. Firstly, the boundary layer flow development was ensured after examining the Y+ value on the blade walls. To study mesh independence, four different mesh models were created by changing the grid density of the inner rotating domain. The number of inflation layer cells (30 layers) and outer domain mesh density were kept constant. Keeping all other parameters constant, the blade was rotated at TSR 3 with time step size of 1° in all three cases. While conducting simulation, careful consideration was given to ensure that the flow around the blade is fully developed. The simulation was performed for 6 complete rotation of the inner domain and the Cm developed by the blade for each degree of rotation was plotted. The flow was considered to be fully developed when the periodic nature of Cm curve is observed for last two rotations. The rotation independence test was also performed for higher TSR rotation. Lastly, the time step size for the transient analysis of VAWT is checked to improve solution accuracy. The time step size is the time required to rotate the inner domain by a certain angle. The smaller the angle rotated for each time step, the precision of the solution was expected to increase. However, computation time increases significantly while doing that. The time step required for the azimuthal increment of $2°, 1°$, and $1/2°$ was selected for the time-independency test. The test was conducted at TSR 3.

3. Results and discussion

3.1 *Results from independence tests*

Initially, the Y+ value on the blade wall is examined to verify that the boundary layer is properly captured. The value of $Y+$ on the blade wall while rotating at TSR3 at 270° azimuth position is shown in **Figure 3**. The value lay well below 1.5, thus confirming that the first layer lies in the viscous

IOP Conf. Series: Materials Science and Engineering **1314** (2024) 012011 doi:10.1088/1757-899X/1314/1/012011

sublayer. For the mesh independency test, the number of mesh, the time taken for one complete rotation of the inner domain, and the average moment coefficient of three different mesh is shown in **Table 3**. **Figure 4**(a) shows the moment coefficient (Cm) for four different mesh models developed by the blade at each degree as it rotates around the axis.The Cm plot of Mesh A shows deviation from other 3 mesh models. It has fluctuating curve while smooth curves are obtained for other models.The plot shows that the Cm developed for other three mesh types are comparable to each other. The average Cm over 360 ° rotation for Mesh C is calculated to be about 3.3 % less than the average Cm of Mesh B. The average Cm of mesh D is 0.86 % less than the average Cm of mesh C. Considering the deviation of results and time taken for the simulation, the mesh type B is selected for further simulations. For time step independence, the plot of Cm vs azimuth angle from the simulation of Mesh B with a time step size of 2°,1°, and 1/2° time step is shown in **Figure 4**(b). The plot shows that the accuracy of the solution could be improved by reducing the time-step size. The average Cm for the 2°,1°, and 1/2° time steps are obtained to be 0.00389, 0.00353, and 0.00335, respectively. The average Cm for a 1° time step is 9.3 % less than the average Cm for a 2° time step, and the average Cm for 1/2 ° is 5.1% less than 1° time-step solutions. It is observed that the error margin of the simulation could be lowered by reducing the time-step size of the simulation. However, the time required for the simulation increases significantly. Hence, the time-step size of 1 ° is selected for the simulation.

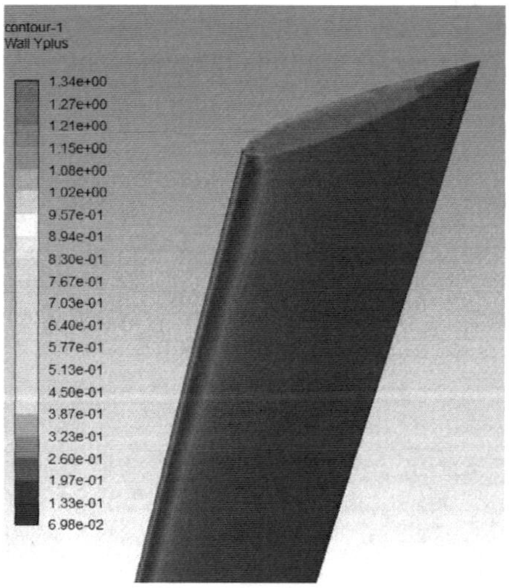

Figure 3. Y+ value on blade walls.

Table 3. Details of cell elements, simulation time, and average Cm for mesh independence test.

Type of mesh	Mesh A	Mesh B	Mesh C	Mesh D
Number of Elements	1,165,137	1,379,455	1,742,688	2,344858
Time taken for one rotation of the inner domain (AMD Ryzen 5800H CPU and 16 GB RAM)	4 hr.	4.4 hr.	5.2 hr.	6 hr.
Average moment coefficient (Cm)	0.00481	0.00365	0.00353	0.00350

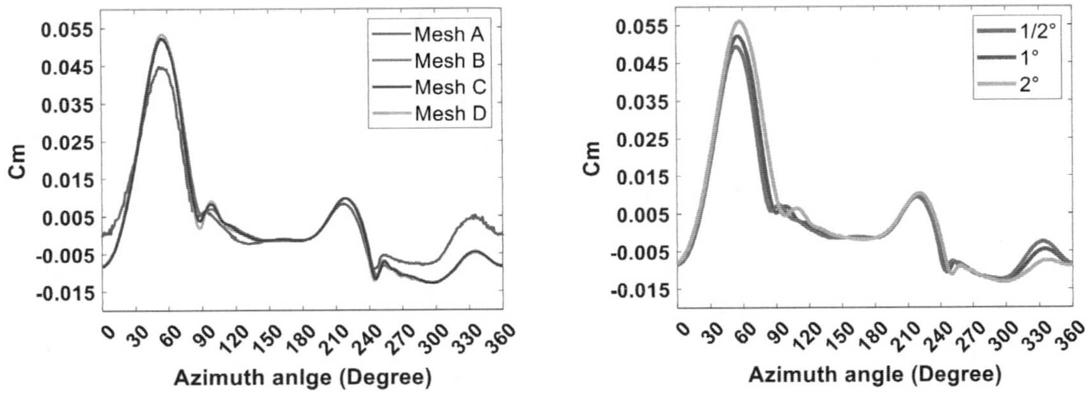

(a) (b)

Figure 4. Moment coefficient (Cm) plot of the blade around 360° rotation for **(a)** Grid convergence test **(b)** Time-step convergence test.

During the mesh and time-step independence test, the inner domain is rotated at least 6 times to allow the flow to be fully developed. This number of rotations required for fully developed flow is verified after comparing the Cm curve for each rotation, as shown in **Figure 5**(a) and (b). The average Cm of Mesh B at TSR 3 with the time-step size of 1° for the 2nd,3rd,$4^{th,}$ and 5th rotation is calculated to be 0.00319,0.00378, 0.00353, and 0.00352 respectively. Likewise, for TSR 4 with the same setup, the average Cm for the 3rd,4th, $5^{th,}$ and 6th rotation is calculated to be 0.01179, 0.01096, 0.01113, and 0.01114, respectively. The calculation of average Cm and analysis of two curves show that the number of rotations should be increased when the rotation speed increases. Accordingly, the inner domain is rotated a higher number of times for higher TSR.

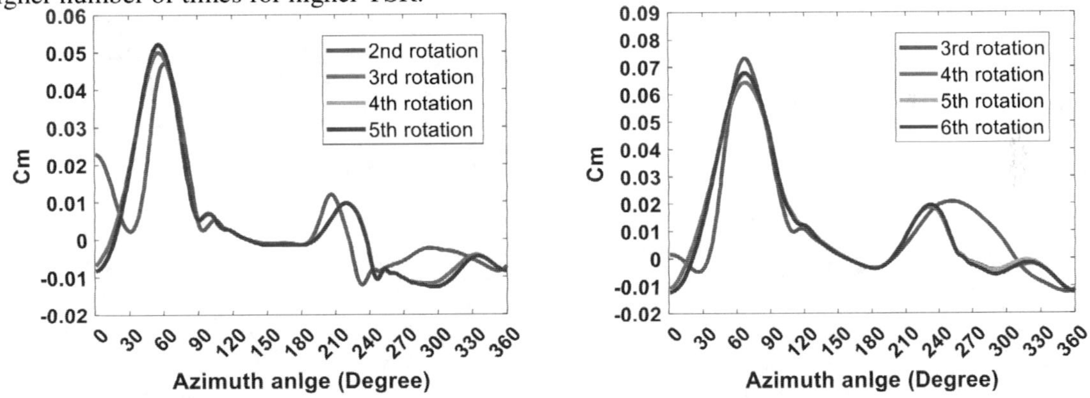

(a) **(b)**

Figure 5. Moment coefficient (Cm) plot of the blade for rotation independency test at **(a)** TSR 3 **(b)** TSR 4.

3.2 *Solution results and interpretation*

The flow developed around the blade for the Mesh B simulation at TSR 3 is visualized using velocity and pressure contours, as shown in **Figure 6**. An XY-plane at a height of 0.6 m from the base of the blade is selected to show the contour at four azimuthal positions of the blade (0°, 90°, 180°, and 270°). The change of velocity and pressure around the blade as the blade rotates is observed in the figure.

IOP Conf. Series: Materials Science and Engineering **1314** (2024) 012011 doi:10.1088/1757-899X/1314/1/012011

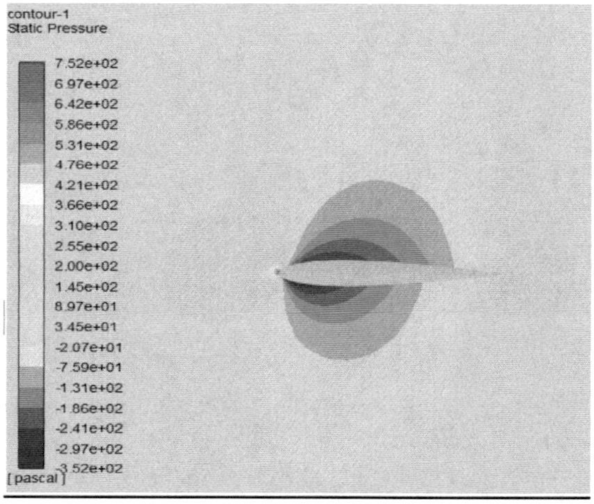

(a) 0°

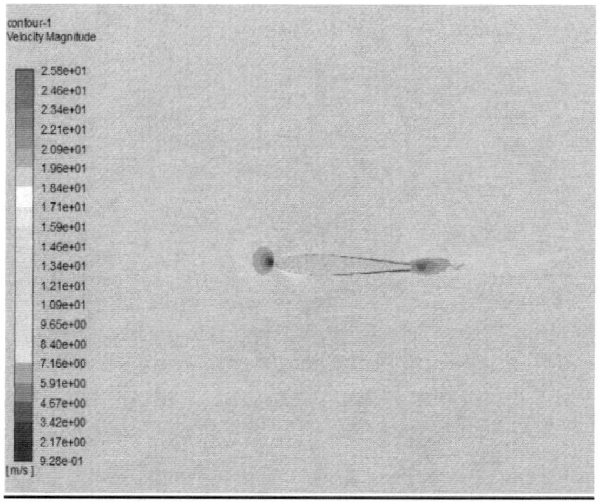

(b)0°

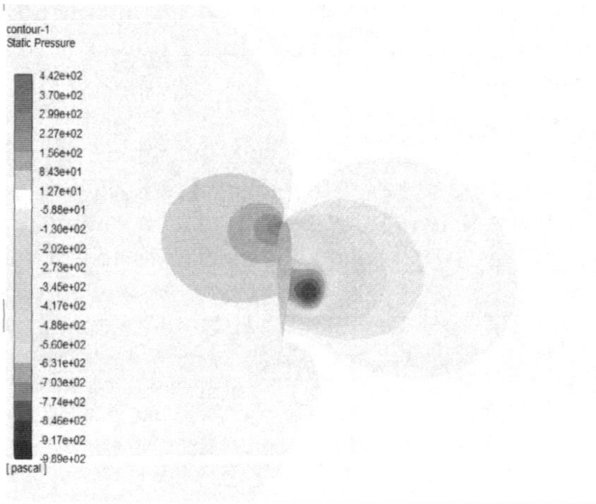

(C)90°

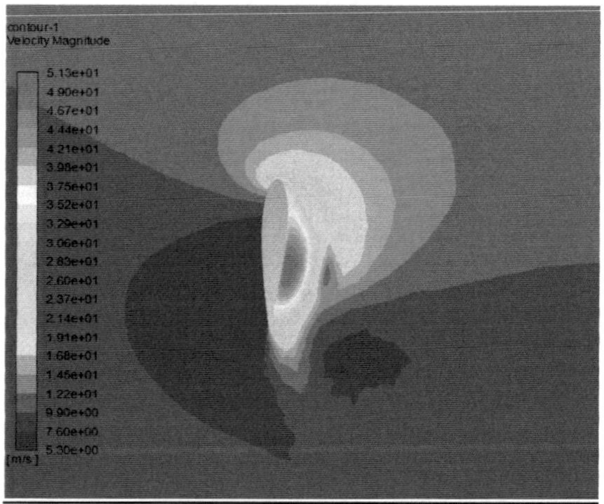

(d)90°

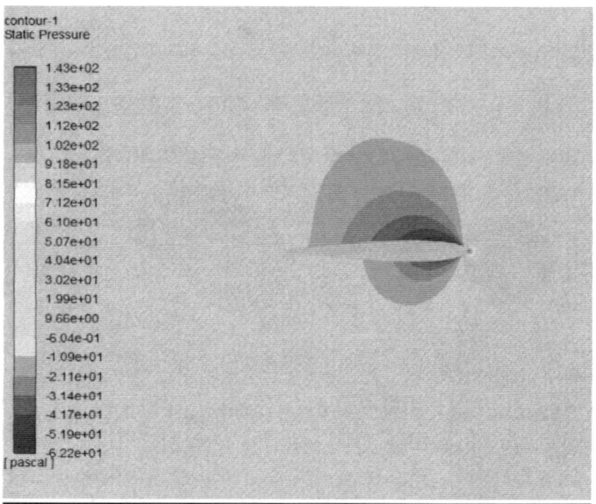

(e) 180°

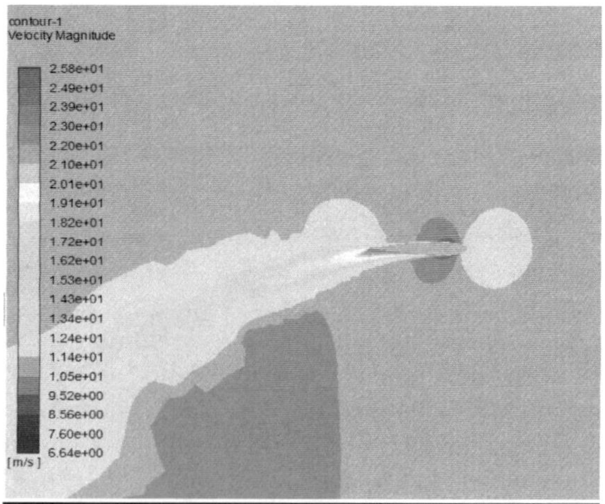

(f) 180 °

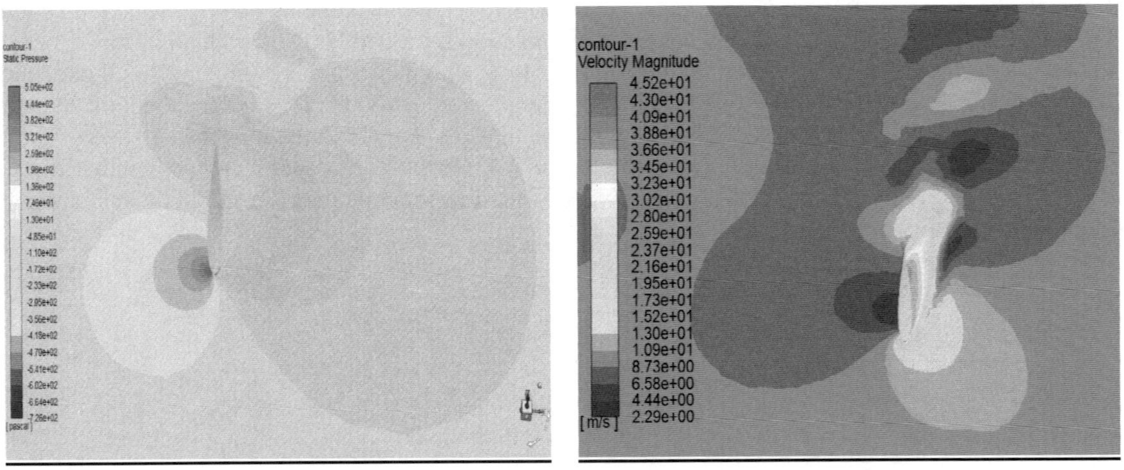

(g) 270 ° **(h) 270 °**

Figure 6. Pressure contour **(a), (c), (e), (g)**and Velocity contour **(b), (d), (f), (h)** around blade at a plane 0.6m above base at different azimuth position.

The solution procedure discussed previously is applied to study the performance of five different blade models. Blade is rotated with a TSR range of 3 to 5 based on previously designed TSR for large-scale V-VAWT in literature(4,11) . The model of the blade is developed by varying the chord length of the base and the tip of the blade. For example, a 20-20 mm blade model is generated with the base and the tip chord length of 20 mm, thus having a constant chord length throughout the blade. Likewise, a tapered blade (125-100 mm blade model) is developed with a base chord length of 125 mm and tip chord length of 100mm, varying chord length throughout the length of the blade. For each model, the blade is rotated at different TSRs. The moment coefficient, from which the total torque and performance coefficient (Cp) provided by the blade is calculated, is plotted on a polar plot for each degree of azimuth angle as shown in **Figure 7**. The polar plot of the lower solidity 20-20 mm blade model in **Figure 7**(a) shows that the blade cannot generate enough positive torque required for the rotation at TSR 3,4 and 5. This is because Cm is too low for most azimuthal positions and even reaches a negative value, lowering the average Cm at TSR 3,4 and 5 to -0.000198, -0.000884, and 0.002069, respectively. Therefore, a 50-50 mm blade model is tested to check if enough torque will be generated. **Figure 7**(b) shows the Cm plot for the 50-50 mm blade model. For this model, the average Cm at TSR 3,4, and 5 is calculated to be 0.00365,0.01096, and 0.018691 respectively. While for a 100-100 mm blade model, the average Cm at TSR 3,4, and 5 is 0.024531, 0.039036, and 0.043313 respectively. The Cm plot for this blade model is shown in **Figure 7**(c). The average Cm increases significantly for the 100-100 mm blade model. So, two different tapered blade models (125-100 mm and 150-100 mm) are compared based on this model. s 7(d) and 7(e) represent these two models' moment plots. It is observed that the average Cm for the 125-100 mm blade model at TSR 3,4 and 5 are 0.040014, 0.031435, and 0.046424, respectively, which is higher than the 100-100 mm blade model. The tapered blade is found to have better aerodynamic performance.

From the polar plot of different blade models, it is visible that the azimuthal position where maximum Cm is achieved changes with increasing TSR and the solidity of the blade. Therefore, the azimuthal position of maximum Cm vs. TSR is plotted for each blade model. It is observed from **Figure 8**(a) that the location and the value of maximum Cm do not increase after the 94 ° azimuth position, even if the solidity of the blade is increased. This suggests the value of Cm saturates at a particular azimuth position while increasing the blade's thickness and taperness. The blade model should be chosen based on this finding. In this paper, the 125-100 mm blade model attained optimum value at 94 ° azimuth position. The results of the 150-100 mm blade model suggest that the aerodynamic performance could not be increased significantly just by increasing the taperness of the blade.

The Performance coefficient (Cp) of each blade model, calculated according to equation 5, is plotted in **Figure 8**(b) to visualize the effect of blade chord length and taperness on aerodynamic performance. The plot suggests that the performance of a thicker blade with a constant chord length is higher than thinner blade with a constant chord length. In the case of the tapered blade, the increment of Cp from the 100-100 mm blade model to the 125-100 mm blade model is significantly greater than the increment of Cp from the 125-100 mm blade model to the 150-100 mm blade model. This shows that the Cp increases if a tapered blade is used, but there is a limitation to the extent Cp could be increased by increasing the taperness of the blade.

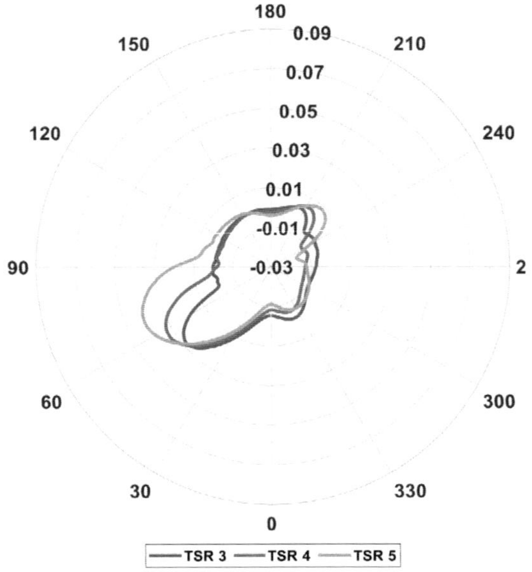

(a) 20-20 mm blade

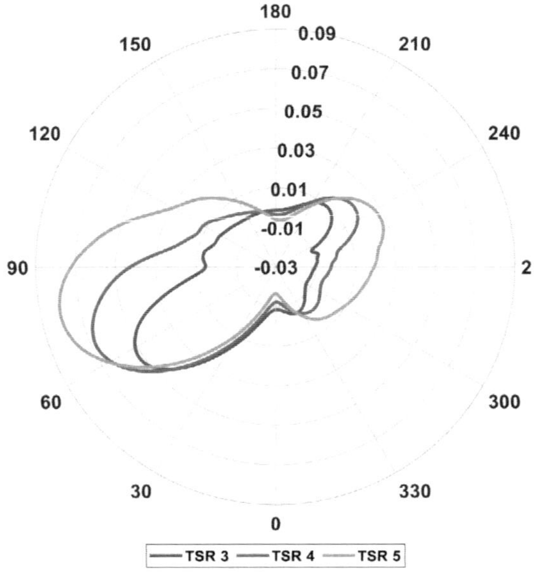

(b) 50-50 mm blade

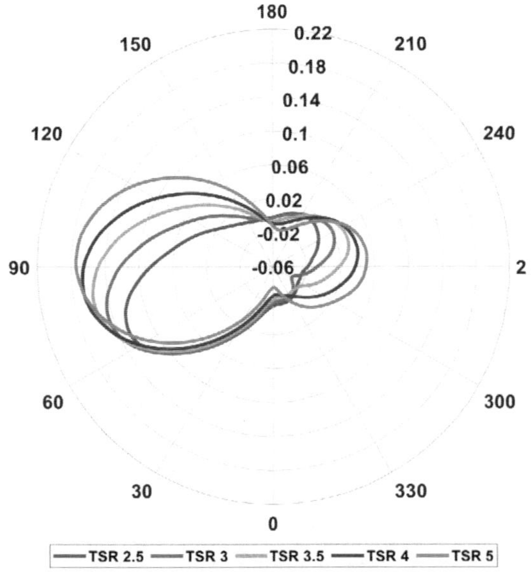

(c) 100-100 mm blade

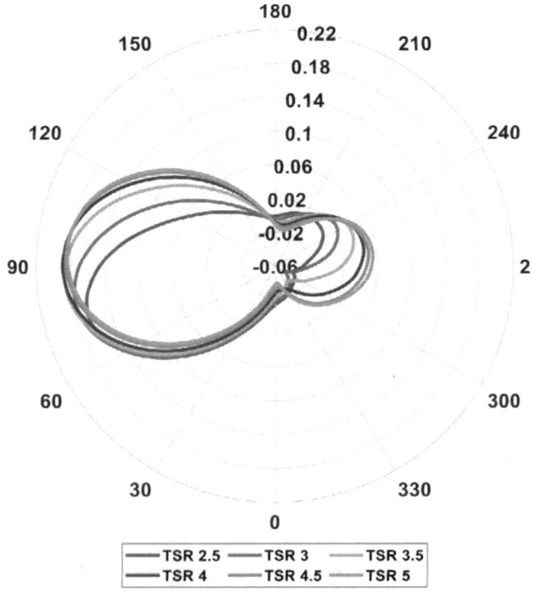

(d) 125-100 mm blade

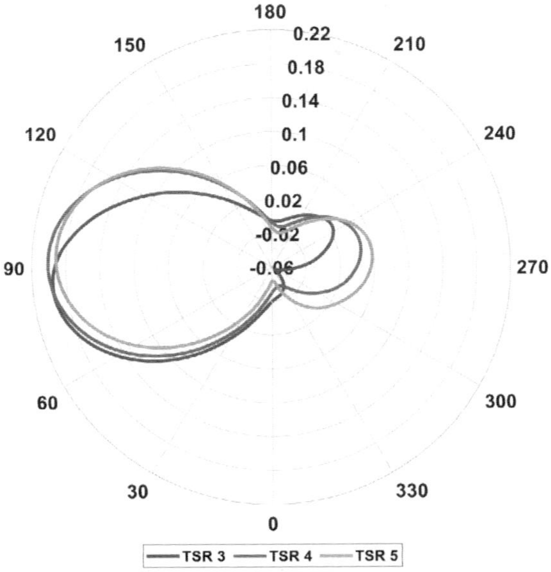

(e) 150-100 mm blade

Figure 7. Polar plot of the moment coefficient (Cm) of different blade models **(a), (b), (c), (d), and (e)** at different TSR.

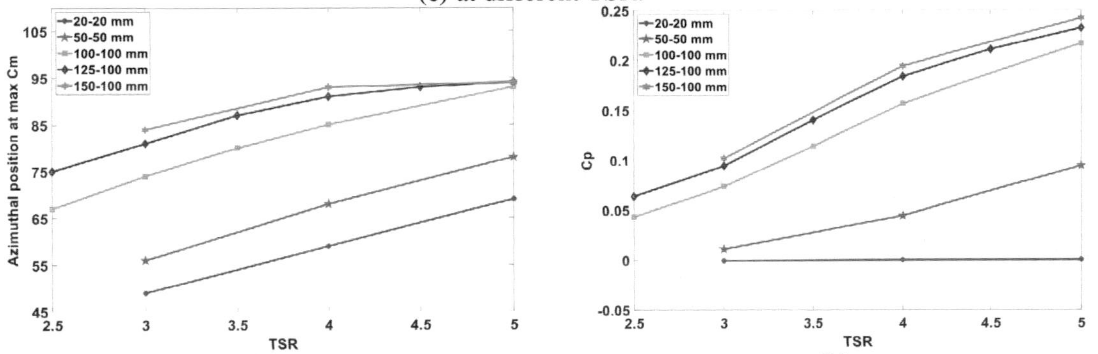

(a) **(b)**

Figure 8. Plot of a different blade models **(a)** azimuth location of maximum Cm vs TSR **(b)** average Cp **vs** TSR.

4. Conclusions

In this paper, 5 different models of blade for the 100 *W* V-shaped VAWT are studied to identify a better-performing blade model that could be used for large-scale V-VAWT in offshore areas. Even though the flow field around the smaller turbine and larger turbine are entirely different, the performance ability of different shaped blades for a smaller turbine is useful to predict the blade model required for the larger turbine.The study's findings suggest the performance of V-VAWT rotor blades is influenced by the design elements of blade shape and chord length. By systematically examining these factors using 3D CFD simulations, optimizing the V-shaped VAWT design to increase energy collection capabilities is possible

Initially, the CFD model having type B mesh and time step size of 1 ° is rotated more than 5 times after condutcing mesh independence, time-step independence, and rotation independence test in the article. After the test, the CFD simulation of different blade models is completed by rotating the blade at varying TSR. The rotation speed is selected based on the designed large-scale VAWT TSR. The 20 mm constant blade model produced a negative average moment coefficient at all TSR. The 50 mm constant blade model produced a positive average moment coefficient of 0.0935 at TSR 5 and can deliver rotational speed. However, the 100 mm blade produced a significantly high average Cp of 0.2167 at TSR 5. Thus, it is selected to study the taperness of the blade. The first tapered blade model, the $125 - 100$ mm blade model, showed a significant increase in aerodynamic performance compared to the 100 mm constant chord model. At TSR 5, the $125 - 100$ mm blade model produced an average Cp of 0.2312. While testing for the $150 - 100$ mm model, the average Cp at TSR 5 is 0.2402. This proved that increasing the taperness of the blade after a certain degree doesn't improve the aerodynamic performance of the blade for the V-VAWT. Therefore, it is concluded that the tapered blade produced better aerodynamic performance than the constant chord blade. However, the taperness of the blade should be checked so that an optimum tapered blade is selected for the turbine.

In future research, it would be valuable to perform experimental verification of CFD results using wind tunnel and compare the $100W$ blade performance with the large-scale blade model.Also, the CFD model could be tested with much higher mesh density and higher blade resolution with the help of powerful computer.

References

1. Tjiu W, Marnoto T, Mat S, Ha M. Darrieus vertical axis wind turbine for power generation I : Assessment of Darrieus VAWT con fi gurations. 2015;75:50–67.
2. Tjiu W, Marnoto T, Mat S, Ha M. Darrieus vertical axis wind turbine for power generation II : Challenges in HAWT and the opportunity of multi-megawatt Darrieus VAWT development. Renew Energy. 2015;75:560–71.
3. Hand B, Cashman A. A review on the historical development of the lift-type vertical axis wind turbine: From onshore to offshore floating application. Sustain Energy Technol Assessments. 2020;38(January).
4. Griffith DT, Barone M, Paquette J, Owens B, Bull D, Simao-Ferriera C, et al. Design Studies for Deep-Water Floating Offshore Vertical Axis Wind Turbines. U.S. Department of Energy. 2018.
5. Renewable Power Generation Costs in 2019 [Internet]. Available from: https://www.irena.org/publications/2020/Jun/Renewable-Power-Costs-in-2019
6. Owens BC. Theoretical Developments and Practical Aspects of Dynamic Systems in Wind Energy Applications. 2013.
7. Sutherland HJ, Berg DE, Ashwill TD. A Retrospective of VAWT Technology. Security. 2012;(January):1–64.
8. Ragni D, Ferreira CS, Correale G. Experimental investigation of an optimized airfoil for vertical-axis wind turbines. Wind Energy. 2015;18(9):1629–43.
9. Baker JR. Features to aid or enable self starting of fixed pitch low solidity vertical axis wind turbines. J Wind Eng Ind Aerodyn. 1983;15(1–3):369–80.
10. Robotham AJ, Sharpe DJ, Taylor DA, Boyle GA. Further Developments in the Taylor "V" Type Vawt Concept. 1986;4:2042–6.
11. Shires A. Development and evaluation of an aerodynamic model for a novel vertical axis wind turbine concept. Energies. 2013;6(5):2501–20.
12. David P, Julia C. Novel offshore vertical axis wind turbines. Cranf Univ.
13. Ferreira CS, Madsen HA, Barone M, Roscher B, Deglaire P, Arduin I. Comparison of aerodynamic models for vertical axis wind turbines. J Phys Conf Ser. 2014;524(1).
14. Hand B, Cashman A, Kelly G. An aerodynamic modelling methodology for an offshore floating vertical axis wind turbine. 2015 Int Conf Renew Energy Res Appl ICRERA 2015. 2015;5:273–7.
15. Chen J, Yang H, Yang M, Xu H, Hu Z. A comprehensive review of the theoretical approaches for the airfoil design of lift-type vertical axis wind turbine. Renew Sustain Energy Rev. 2015;51:1709–20.
16. Moghimi M, Motawej H. Developed DMST model for performance analysis and parametric evaluation of Gorlov vertical axis wind turbines. Sustain Energy Technol Assessments. 2020;37(October 2019):100616.

17. Bangga G, Dessoky A, Lutz T, Kr E. Improved double-multiple-streamtube approach for H-Darrieus vertical axis wind turbine computations. Energy. 2019;182:673–88.
18. Soraghan CE. Aerodynamic Modelling and Control of Vertical Axis Wind Turbines. Ph.D Thesis. 2014.
19. Almohammadi KM, Ingham DB, Ma L, Pourkashan M. Computational fluid dynamics (CFD) mesh independency techniques for a straight blade vertical axis wind turbine. Energy. 2013;58:483–93.
20. Hand B, Kelly G, Cashman A. Numerical simulation of a vertical axis wind turbine airfoil experiencing dynamic stall at high Reynolds numbers. Comput Fluids. 2017;149:12–30.
21. Rezaeiha A, Montazeri H, Blocken B. On the accuracy of turbulence models for CFD simulations of vertical axis wind turbines. Energy. 2019;180:838–57.
22. Bangga G, Dessoky A, Wu Z, Rogowski K, Hansen MOL. Accuracy and consistency of CFD and engineering models for simulating vertical axis wind turbine loads. Energy. 2020;206:118087.
23. Howell R, Qin N, Edwards J, Durrani N. Wind tunnel and numerical study of a small vertical axis wind turbine. Renew Energy. 2010;35(2):412–22.
24. Zhang LX, Liang YB, Liu XH, Jiao QF, Guo J. Aerodynamic performance prediction of straight-bladed vertical axis wind turbine based on CFD. Adv Mech Eng. 2013;2013.
25. Alaimo A, Esposito A, Messineo A, Orlando C, Tumino D. 3D CFD analysis of a vertical axis wind turbine. Energies. 2015;8(4):3013–33.
26. ANSYS I. ANSYS FLUENT Tutorial Guide 17.0. 2016;15317(January):724–46.
27. Rezaeiha A, Montazeri H, Blocken B. On the accuracy of turbulence models for CFD simulations of vertical axis wind turbines. Energy. 2019;180:838–57.

Comparative analysis of biomass boiler, heat pump, direct electric and solar-biomass based heating systems for small residential building in Kathmandu

N S Timilsina[1,2*], N Adhikari[1,2], S Gautam[2], A Aacharya[1,2], S Satyal[2], P Pandey[1,2], B Baral[1,2]

[1]Energy Systems and Technology Research Laboratory (ESTRL), Kathmandu University, Nepal

[2]Department of Mechanical Engineering, Kathmandu University, Nepal

*Corresponding author E-mail: timilsina.nirazan@gmail.com

Abstract. Globally, 42 % of the energy produced is used for heating and cooling applications to maintain thermal comfort for the human body. The selection of these heating systems is governed by initial investment and operating expenses. This study presents the technical and economic assessment of four heating systems: a biomass-based radiator system, an air-air heat pump system, a direct electric heater system, and a hybrid-biomass solar-based radiator system. The heating requirement of the building of an area of 94 m^2 for the winter months of Kathmandu was determined. Later, the heating systems of standard size were selected for technical and economic assessment to choose the suitable one. Based on degree-day method, the annual heating requirement of the building was estimated to be 10890.5 kWh to maintain a temperature of 20°C. The total yearly energy of 10880 kWh, 3203 kWh, 10890.5 kWh, and 6013.17 kWh were consumed for the biomass, the heat pump, the direct electric, and the hybrid system, respectively. The equivalent energy costs for operation were estimated to be Rs 52,500, Rs 38,436, Rs. 1,19,795, and Rs. 29,000 for the respective systems. Considering the initial and operating costs over 15 years time period, the present worth of the hybrid system is the largest amongst all other systems. Hence, from the study, the hybrid biomass- solar-based radiator is the best heating system for Kathmandu, considering the energy costs and NPW.

Keywords: Energy consumption, Heating systems, Economic Analysis, Degree Day method, and Solar-Biomass Boiler

1. Introduction

It is estimated that about 46 % of the heat produced is consumed by buildings for space heating [1]. Traditionally, the demand for space heating is fulfilled by central heating systems in colder climates and by electric heaters in moderately cold climates. Electricity, biomass, coal and natural gas are the commonly used fuel sources for generating heat [1]. In Nepal, a variety of climatic conditions exists - ranging from 24°C in the southern Terai (plainland) to sub-zero climate in the northern Himalayan mountains [2]. About 84 % of the energy is consumed in residential homes [3] and 77% of the total energy is supplied by biomass in Nepal [4]. Amongst the various types of biomasses available such as cow dung, firewood, and agricultural residues, firewood is one of the most used energy sources [5]. In a recent survey by the Central Bureau of Statistics, the average number of members in Nepalese households is just above 4 in urban areas and about 50% of the people living in urban areas live in Kathmandu, where a typical house has concrete roofs and brick wall [5].

Central heating systems generally have a boiler or furnace in a single location commonly known as the mechanical room, where heat is generated, and the generated heat is distributed throughout the building through pipes or ducts networks. Generally, hot water or steam generated by the boiler is supplied in the pipes and heat is transferred via radiators, whereas ducts are responsible for supplying hot air generated by the furnace. The boilers or furnaces require an energy source to generate heat. Commonly, electricity, firewood, biomass, coal, and natural gas are used as fuel sources for generating heat; and radiators are used to disperse the heat into rooms.

Air-conditioning is gaining popularity in Nepal with the total imports averaging about NRs 2.251 billion over 5 years increasing at an average rate of 13 % [6]. For the climatic conditions of Nepal with summers and winters, heat pumps offer both heating and cooling solutions with optimum efficiency under both conditions. A central variable refrigerant flow (VRF) based system efficiently pumps the refrigerant to the split units depending upon the demand. These systems have an outdoor unit, which contains a compressor, evaporator, and indoor units for heating/cooling applications. Direct electric heating is another popular source of heating in Nepal with imports valued at an average of NRs 241 million over the last five years [6]. Direct electric heating systems come in wide varieties ranging from halogen heaters to nichrome wires to convectors in the Nepalese market. As the electricity passes through a high resistance wire, these systems convert the electric energy to heat.

Renewable energy sources such as solar energy can be beneficial as Kathmandu gets about 4.7 kWh/m^2 of solar irradiance and the solar radiation for Kathmandu is 777 W/m^2 and 300 sunny days per year [7, 8, 9]. The selection of a proper heating system is determined by the initial cost and the day-to-day operation costs. Additionally, the efficiency of a system plays a pivotal role in determining which system to choose as efficient models tend to be expensive in the initial phase while the operating costs are nominal [10]. In this study, the heating systems are evaluated based on energy consumption and the total economic value.

2. Methodology

2.1 System Description

The analysis was carried out for a house in Kathmandu consisting of two bedrooms, one hall, and one kitchen (2BHK) with an area of 94 m^2, assuming no shadowing effect from other houses. The building has a concrete floor and ceiling with brick walls for outer cover and partition. A layer of plaster is assumed to be filled in the floor, whereas all the walls and the ceiling consist of a double layer of plaster. The living room, bedrooms and kitchen are the rooms, where occupants reside so thermal comfort is essential in these rooms. The thermal comfort conditions were calculated for indoor conditions: temperature of 20 °C and relative humidity of 50 % [11]. Also, the temperature of Kathmandu was assumed for the whole study as it was the only available data for Nepal.

The heating load is the energy required to size the heating systems to a pre-defined temperature for the worst temperature conditions. The material properties determine the overall heat transfer coefficient or the U-value as in Table 1, which is then used to calculate the heating load as in Table 3. The U-value specifies how well heat is transferred over the medium [11].

IOP Conf. Series: Materials Science and Engineering **1314** (2024) 012012 doi:10.1088/1757-899X/1314/1/012012

$$\frac{1}{U} = \frac{1}{h_o} + \frac{l}{k} + \frac{1}{h_i} \tag{1}$$

$$Q_{DL} = U\,A\,dT + \frac{V}{SV}(H_i - H_o) \tag{2}$$

Where l- Thickness of material [m], k- Thermal conductivity of material [W/mK], h_o- Convective heat transfer coefficient of outside air [W/m^{2}K], h_i - Convective heat transfer coefficient of inside air [W/m^{2}K], U – Overall Heat transfer Coefficient [W/m^{2}K], A - Area of the whole building envelope [m^2], dT- Temperature difference between room and outside air/rooms [°C], V- Volume of air in the room, SV- Specific volume of outside air, $(H_i - H_o)$- Difference in enthalpies of inside and outside air [kJ/kg], Q_{DL} – Design Heating Load [kW]

Table 1. U values of building components

Item	Material	Thickness [in]	U value [W/m^{2}K]
Walls	Plaster-Brick-Plaster	0.2 – 9 - 0.2	2.383
Roof	Plaster-Concrete-Plaster	1 - 6 - 1	2.68
Floor	Plaster-Concrete	1 - 8	3.13
Windows	Glass	0.2	3.30
Door	Wood	1.2	1.98 (outside) 1.78 (inside)

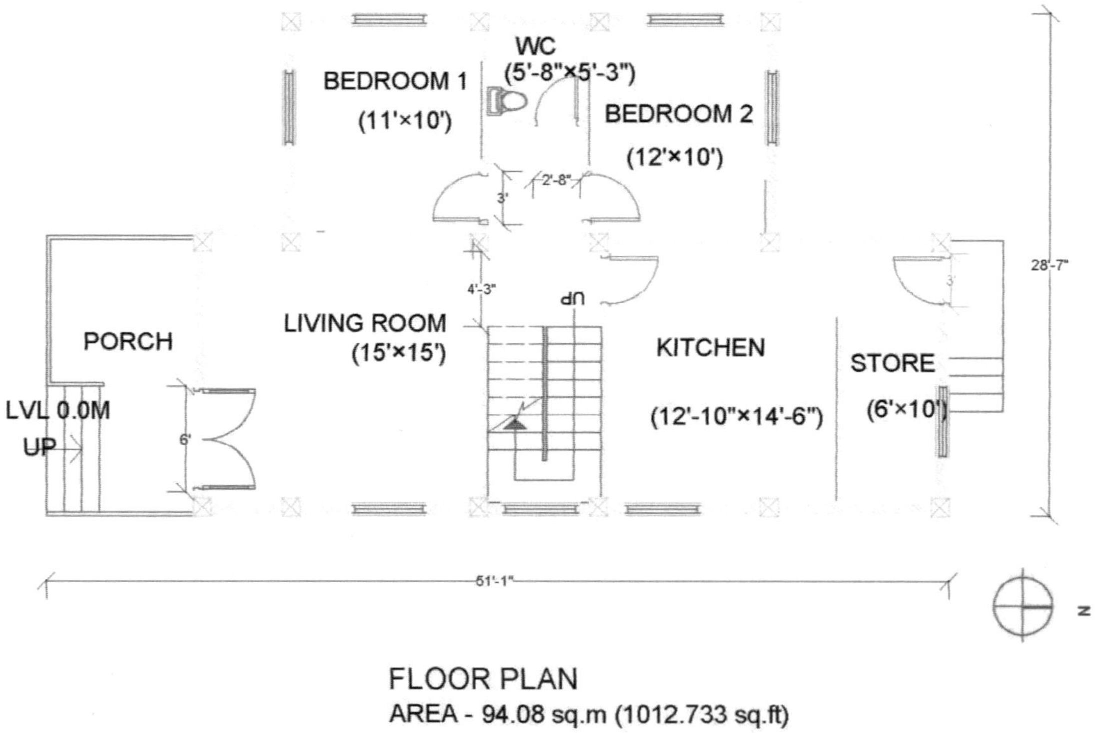

Figure 1. Schematic of building layout

2.2 Energy Consumption Modelling

The heating load obtained using Equation 2 is used for the sizing of the equipment. For daily energy comparison, an hourly temperature profile of Kathmandu is used to calculate the hourly heat load. The building is assumed to be completely air-tight and no leakages are present. Thus, the ventilation and infiltration losses are neglected in the calculation. So, for the hourly comparison, the hourly heating load can be calculated for the hourly temperature profile as

$$Q_{HL} = (U \times A)_{building} \times dT \quad (3)$$

Where, Q_{HL}- Heating load [kWh]; U – Overall heat transfer coefficient of whole building [W/m^{2}K]; A- Area of the whole building envelope [m^2]; $dT = T_{room} - T_{ambient}$ – Temperature difference between indoors and outdoors [°C]

Also, the seasonal heating load can be calculated by using the degree day method, which gives an accurate estimate of the annual energy consumption, where the heating systems operate for the whole season. It is an integral function of time which varies with temperature. A balance temperature of 18.3°C is considered at which the heat gain and loss in the building is the same [11].

The values of heating degree days were calculated for the winter months from November to February as in Table 2 by using Equation 5.

$$Q_{seasonal} = (UA)_{building} \times HDD_{seasonal} \times 24 \quad (4)$$

$$HDD_{seasonal} = \sum N_{days} \times (18.3 - T_{avg}) \quad (5)$$

Where, $Q_{seasonal}$ – Seasonal Heating Load [kWh]; $(UA)_{building}$ – Overall heat transfer coefficient of building & its area [kW/°C]; 24 – Conversion of day into hour; T_{avg} – Average temperature of the day [°C].

Table 2 Heating Degree Days for Kathmandu [12]

Month	Tavg [°C]	HDD [°C-day]
November	14.80	104.9
December	13.13	222.7
January	11.12	222.6
February	13.13	129.3

In January, the recorded temperature was the lowest in the winter months as in Table 2. The temperature profile on 4th January 2022 was taken for the calculation of daily energy consumption in this study. Normally, heating/cooling systems are designed for the worst-case scenario by considering the minimum temperature for load calculation and sizing [11]. For this study, the minimum temperature was considered for the analysis, which was experienced on 4th January 2022 with a wide range of temperature variation from 5.5°C to 17 °C.

IOP Conf. Series: Materials Science and Engineering **1314** (2024) 012012 doi:10.1088/1757-899X/1314/1/012012

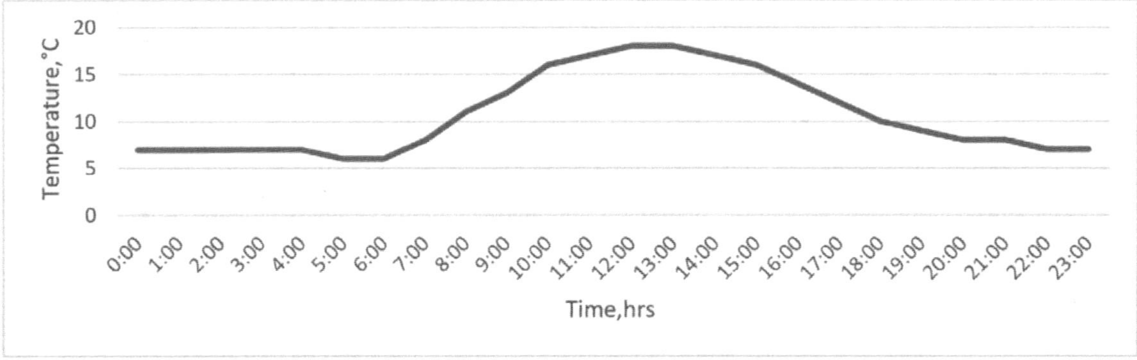

Figure 2. Hourly temperature of Kathmandu on January 4, 2022 [13]

2.3 Sizing of Heating Systems

The selection of the heating systems was done according to the estimated heating requirements of each room using Equation 2 as in Table 3. Four different heating systems were selected for the study. Since biomass is the dominant fuel source in Nepal with firewood being the extensively used biomass source, Biomass-based boiler was selected to operate for firewood. Also, since hot water systems are easy to install, the study focused on a biomass-boiler based hot water radiator heating system. The other three heating systems were: air-to-air heat pump system, a direct electric heater system and a hybrid-biomass solar-based radiator system used for comparison. The air-to-air heat pump system was taken into consideration as it offers the option of both cooling and heating and therefore is used widely in commercial as well as residential buildings in Nepal. The direct electric heater system is another commonly used heating solution in Nepal, while the hybrid solar-biomass system used two renewable technologies together. In this study, solar-biomass-based hybrid systems were considered for comparison due to their future potential as heating systems.

Table 3. Heating load calculation for whole building

Room	**Heating Load, kW**
Living Room	4.04
Bedroom 1 (adjacent to Living Room)	2.24
Kitchen	2.84
Bedroom 2 (adjacent to Kitchen)	2.41
Total Building	**11.541**

2.3.1 Biomass-based radiator system. The biomass-boiler based hot water radiator heating system comprises of radiators, which distributes the heated water from the boiler to the rooms. A 12-kW boiler having 95% boiler efficiency [15] was used along with copper pipes wrapped around insulation to supply the hot water to the rooms. Also, a circulating motor, an expansion buffer tank and various valves are fitted to control the flow of hot water and for safety concerns, as shown in Figure 3. The summary of system size is shown in Table 4. The system uses firewood as the source of fuel with a calorific value of 14000 kJ/kg [16]. The energy consumed by the system for daily and seasonal usage is given as

$$m_{fuel} = \frac{\dot{Q}_{HL}\,3600}{\eta_{boiler}\,CV_{fuel}} \; and \; \frac{Q_{seasonal}\,3600}{\eta_{boiler}\,CV_{fuel}} \tag{6}$$

Where η_{boiler}- Efficiency of boiler, m_{fuel} – Mass of fuel consumed [kg], CV_{fuel} – Calorific value of fuel [KJ/kg]

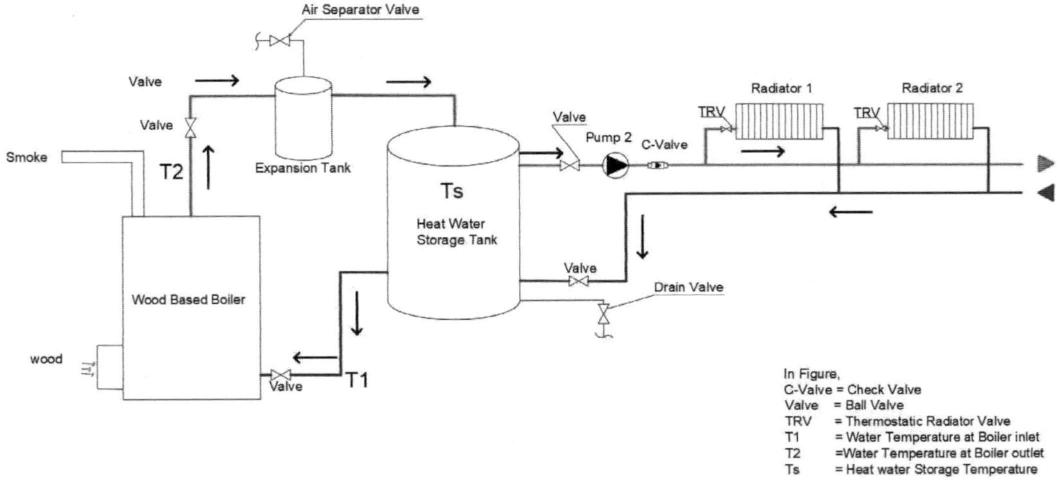

Figure 3. Schematic diagram of biomass-based heating system

2.3.2 Air-Air heat pump system. The air-to-air heat pump system comprises of an outdoor unit and individual indoor units for each room. Based on calculated heating demand, outdoor unit of Daikin RXYRQ5ARV16 of heating capacity of 16 kW with a COP 3.4 was selected, since these heat pumps are commonly used in Nepal [17]. The indoor units were sized using the product catalogue as depicted in Table 4 [18]. The installation of the heat pump also comprises of copper pipes of variable diameters with insulation, drain pipes and the need of inserting the refrigerant gas. The system uses single phase electricity to operate. The energy consumed by the heat pump for daily and seasonal use is given as:

$$Energy\ Consumed\ by\ Heat\ Pump = \frac{\dot{Q}_{HL}}{COP}\ and\ \frac{\dot{Q}_{seasonal}}{COP} \tag{7}$$

2.3.3 Direct electric heater system. The direct electric heater system used in this study comprises heaters having 2 kW and 1 kW rated power output, which are installed in configurations mentioned in Table 4. A glass convector-type heater was considered for this study. Although some systems include temperature controlling thermostat and timers, such heaters are not available in the Nepalese market. Since, these heaters use 100 % of the electricity to convert it into heat energy, the energy balance for the system is given respectively for daily and seasonal use is given as:

$$Energy\ Consumed\ by\ Heater = Q_{HL}\ and\ Q_{seasonal} \tag{8}$$

2.3.4 Hybrid-biomass solar-based radiator system. The hybrid system was designed to utilize the solar energy available during the day from about 10:00 am to 4:00 pm. The system adds a solar circuit to the existing setup of the biomass-based radiator system. In sun hours, the operation of the boiler is halted and solar circuit will operate to supply the hot water to the radiators. The solar circuit consists of a flat plate collector of 12 m^2 area with a 350 liters extra water storage tank. The energy consumed for a season of heating is calculated by subtracting by a factor of 35 kW to the number of days of operation (N) as:

$$m_{fuel-solar} = \frac{(\dot{Q}_{seasonal} - 35 \times N) \times 36}{\eta_{boiler} \times CV_{fuel}} \tag{9}$$

IOP Conf. Series: Materials Science and Engineering **1314** (2024) 012012 doi:10.1088/1757-899X/1314/1/012012

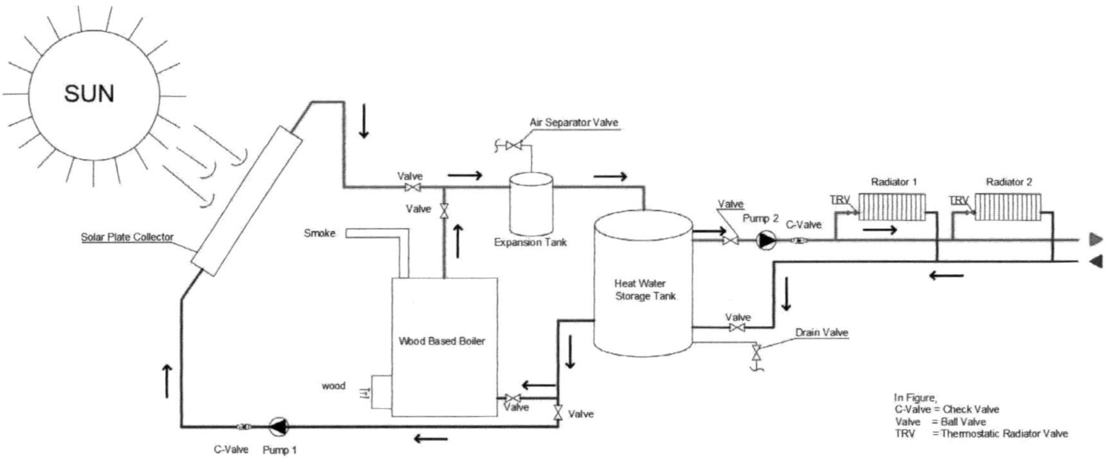

Figure 4. Schematic diagram of hybrid system

Table 4. System sizing

Room	Design Load	Heat Pump Indoor units		Biomass System Radiator		Direct Electric Heater		Hybrid System
	[kW]	[kW]	[nos.]	[W]	[nos.]	[kW]	[nos.]	Radiator
Bedroom 1	2.24	3.2	1	2503	1	2 & 1	1 & 1	Same as the
Bedroom 2	2.42	3.2	1	2503	1	2 & 1	1 & 1	Biomass
Kitchen	2.84	4	1	2861	1	2 & 1	1 & 1	system
Livingroom	4.04	5	1	2146	2	2	2	
Total	**11.54**	**15.4**	**4**	**12,159**	**5**	**13**	**8**	

Table 4 summarizes the energy requirement for each room of the building with the total number of selected heating components and power rating. The possible arrangements of these systems in the layout are shown in Figure 4(a-c). In total, four indoor units of the heat pump system (one in each room with different capacities) were arranged in the building. Other systems of equivalent capacity were also distributed to meet the heating demand of the building. The energy analysis was conducted for these arrangements of each system in the building.

IOP Conf. Series: Materials Science and Engineering **1314** (2024) 012012 doi:10.1088/1757-899X/1314/1/012012

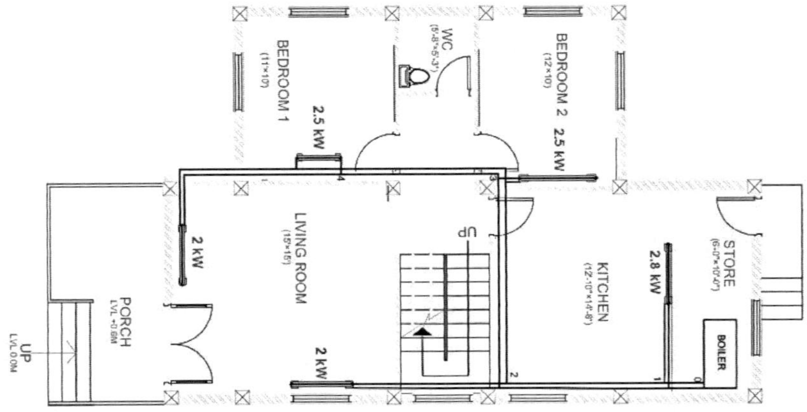

(a)

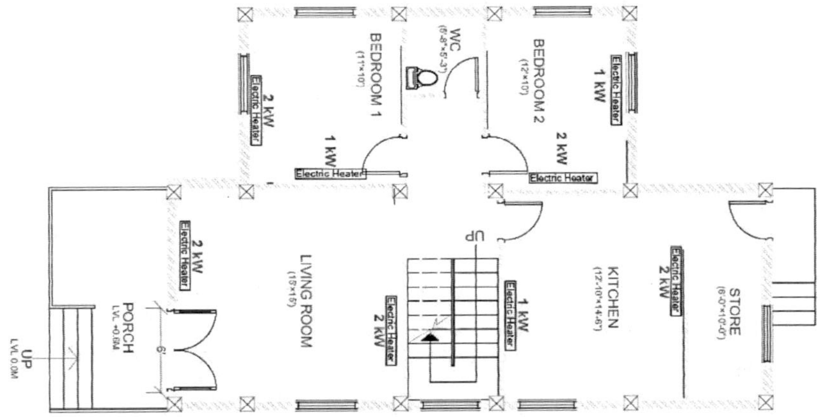

(b)

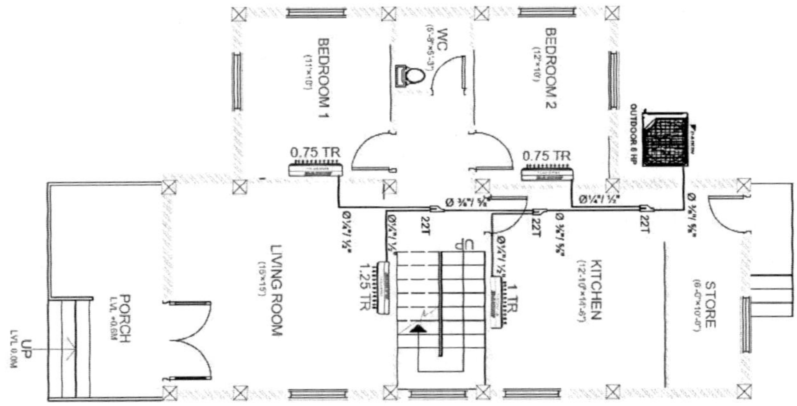

(c)

Figure 5. Floor Plan Layout of **(a)** Radiator System **(b)** Direct Electric System **(c)** Heat Pump System

2.4 Economic Analysis

The economic analysis of the system was done by the net present worth analysis method. It is an equivalence method in which all project's cash flows including the future are converted into a single present value by using a rate known as the minimum attractive rate of return (MARR). The least negative or the highest positive value of the present worth is chosen as the best project while comparing the present worth of various projects [14]. The NPW is calculated by equation 6.

$$NPW = \frac{Cash\ In - Cash\ Out}{(1+R)^t} - \ Initial\ Investment \quad (10)$$

Where, R = MARR, t = Time Period

The initial costs of the various systems were derived after extensive market surveys conducted in Kathmandu, Nepal. Quotations from various HVAC suppliers were used to assess the specifications and total cost to estimate the initial costs. The initial cost of the biomass-based radiator systems, the air-air heat pump system, direct electric system and hybrid system was estimated to be NRs 350,000, NRs 650,000, NRs 1,19,795, and NRs 455,700, respectively.

3. Results and Discussions

3.1 Energy Consumption and Costs

The study considered the rate of electricity as NRs 11 per kWh for single-phase electricity and Rs 15 per kg for firewood and the 24-hour operation of the systems for the whole season. The operating costs were summarized in Figure 5. Equations 3, 4, 7, 8, 9, and 10 were used to calculate the values.

For the temperature profile of Kathmandu, the heat pump consumed less energy than other systems due to its high COP and consumed the least amount of energy. It was found that this system consumed about 42kWh of energy daily and 3203 kWh of electricity in the whole season which amounts to NRs 504 and NRs 38,436 for daily and seasonal operation.

The biomass boiler should have been cheaper to operate than the heat pump but the low calorific value of the firewood available in the Nepalese market causes the system to consume more energy [16]. The daily and seasonal energy consumption was estimated to be 123.5 kg and 10,880 kg of firewood costing NRs 595.5 and NRs 52,500 respectively. The difference between the operation costs of the heat pump and the biomass boiler-based systems was almost Rs 14,000. This finding concurred with the finding of Olkowski et al., which found heat pumps to be cheaper to operate than biomass systems by USD 130 even in the climate of Poland [10]. Though climatic condition differs in Nepal and Poland, the figures do not match with each other but still it aligned with the estimation from this study.

The direct electric heater system was found to be the most uneconomical system, consuming about 142.2 kWh and 10890.5 kWh of electricity, which cost NRs 1564.20 and NRs 119,795.5 respectively for the daily and seasonal operation.

The hybrid system utilizes the total potential of the sun. On a bright sunny day, the hybrid system was found to be effective in reducing the consumption of firewood. On daily operation, it consumed only about 100kg of firewood costing NRs 480, which was even lower than the operation of the heat pump. Seasonally, the hybrid system consumed about 6013.7 kg of firewood amounting to NRs 29,000 in operation. Hence, the hybrid biomass solar based system was the most energy efficient system.

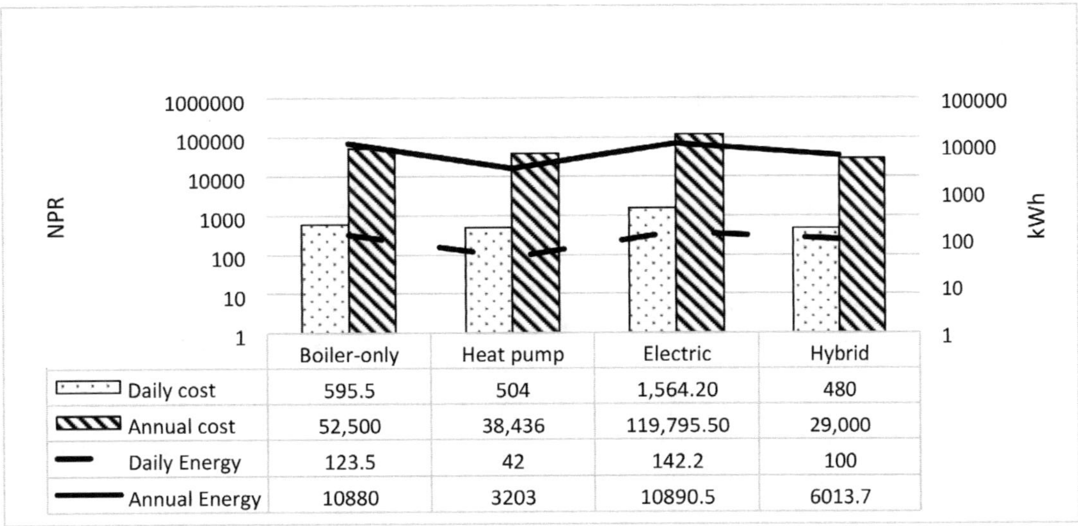

Figure 6. Energy and cost comparison of the systems

3.2 Net Present Worth Analysis (NPW)

The net present values of the systems were calculated based on the annual operating costs derived from the annual costs as in Figure 5. The salvage value was derived through an annual 10% depreciation for 10 years. The minimum annual rate of return was assumed as 10% and the NPW was calculated for a period of 15 years. The direct electric heaters were assumed to be replaced in about 7.5 years. The NPW was calculated as in Table 5.

The calculated values of NPW were negative due to the comparatively minimum salvage value of the systems. Furthermore, no income was generated by the systems and all the costs associated with the systems were expenses. Although the heaters have the least investment upfront, they were expensive to operate and therefore have the least NPW. Meanwhile, the high initial cost of the heat pump system causes it to have a lower NPW than the boiler-only system despite having lower operating costs. The hybrid-biomass solar-based system has the least negative present worth, therefore, it is the best investment out of all the various systems.

Table 5. Present Worth Comparison of the various systems

Parameters	Heat Pump	Hybrid	Boiler-only	Direct Electric
Initial Cost, NRs	6,50,000	4,55,700	3,50,000	90,000
Annual Operating cost (AOC), NPR/year	38,437	29,000	52,510	1,19,795
Salvage value S, NPR	65,000	45,570	35,000	9,000
Life, years	15	15	15	7.5
MARR	10%	10%	10%	10%
NPW	**-926,794**	**-665,367**	**-741,016**	**-1,038,647**

4. Conclusion

Four different heating systems were compared for energy consumption based on energy balance and NPW to find out the appropriate heating system for the temperature profile of Kathmandu. This study concludes the following:

- Hybrid system was found to be the cheapest to operate in considering both the daily and annual energy consumption and was the best investment for over a period of 15 years. Since, the study was carried out considering fully sunny days, the system may consume more energy in cloudy days.
- The heat pump system was found to be the second most suitable option. Since the whole study does not account for operation in the summers, the heat pump system might be the best system in such case.
- The low calorific value of the firewood causes the biomass system to be slightly more expensive than the heat pump system. Also, a lower temperature profile causes the heat pump to lose its COP, and in such cases, the biomass system may be effective.
- The direct electric heater system was found to be the most energy-intensive and expensive system among the four systems.

5. References

[1] IEA 2019 *Renewables 2019*.(https://www.iea.org/reports/renewables-2019)
[2] Asian Development Bank and World Bank 2021 *Climate Risk Country Profile: Nepal* (https://www.adb.org/publications/climate-risk-country-profile-nepal)
[3] Asian Development Bank 2017 *Nepal Energy Sector Assessment, Strategy, and Road Map* (https://www.adb.org/publications/nepal-energy-strategy-roadmap.)
[4] Ministry of Population and Environment, Government of Nepal 2017 *Biomass Energy Strategy 2017* (Kathmandu)
[5] Central Bureau of Statistics, Government of Nepal and UNDP 2016 *ANNUAL HOUSEHOLD SURVEY 2015/16 (MAJOR FINDINGS)* (Kathmandu: Central Bureau of Statistics, Government of Nepal).
[6] Department of Customs, Government of Nepal *Statistics FY2079/80, FY2078/79, FY2077/78, FY2076/77 and FY2075/76* (https://www.customs.gov.np/page/statistics)..
[7] Poudyal K N, Bhattarai B K, Sapkota B, Kjeldstad B and Daponte P 2013 *Estimation of The Daily Global Solar Radiation; Nepal Experience* (Measurement vol. 46 no. 6) pp. 1807-1817.
[8] Water and Energy Commission Secretariat 2010 *Energy Sector Synopsis Report Nepal 20092010* (WECS).
[9] Poudyal K N, Bhattarai B K, Sapkota B, and Kjeldstad B 2012 *Solar Radiation Potential at Four Sites of Nepal* (Journal of Institute of Engineering vol. 8, no. 3) pp. 189-197.
[10] Olkowski T, Lipiński S and Olędzka A 2017 *Comparative analysis of heat pump and biomass boiler for small detached house heating* (Polanica Zdrój : Poland / International Conference Energy, Environment and Material Systems).
[11] ASHRAE 1997 *ASHRAE Handbook Fundamentals, SI Edition*
[12] National Center for Environmental Information 2022 *Climate Data Online* (National Oceanic and Atmospheric Administration, US Government).
[13] TimeandDate *Hour-by-Hour Forecast for Kathmandu, Nepal*

(https://www.timeanddate.com/weather/nepal/kathmandu/hourly.).
[14] C. Park 2004 *Fundamental of Engineering Economics* (Pearson).
[15] Balkan Energy *Pellet boiler Eco Spar Minima B, 12kW.*(https://balkanenergy.net/product/ecospar-minima-b-12kw)
[16] Bhatt B P, Sarangi S K and De L C 2010 *Fuelwood Characteristics of Some Firewood Trees and Shrubs of Eastern Himalaya, India* (Energy Sources, Part A: Recovery, Utilization, and Environmental Effects, vol. 32, no. 5) pp. 469-474.
[17] Daikin Airconditioning *Daikin VRV Catalogue* (India.)
[18] Daikin Airconditioning India *VRV Indoor Units* (https://www.vrvhome.com/vrv-indoor-unit.)

Acknowledgments
The authors would like to thank Energy Systems and Technology Research Laboratory (ESTRL), Kathmandu University for providing a workspace, opportunity and support to conduct this research in our final year of undergraduate study. Also, we would like to express our sincere gratitude to Miss Sanjeena Jha for her support in the design of the floor plan of the building.

IOP Conf. Series: Materials Science and Engineering **1314** (2024) 012013 doi:10.1088/1757-899X/1314/1/012013

Techno-Economic Feasibility Study of Solar Powered Rural Community Heat Grid System in Langtang

M Ghimire[1], P Dahal[1], S Bhandari[2], R Kandel[2], Y Niroula[1], S Pandit[1], S Shrestha[1] and M Shah[1]

[1] Department of Mechanical Engineering, Kathmandu University, Dhulikhel, Nepal

[2] Department of Environmental Science and Engineering, Kathmandu University, Dhulikhel, Nepal

malesh.shah@ku.edu.np

Abstract. Langtang, a popular tourist destination in Nepal's northern Himalayas, relies heavily on solid fuels like coal, wood, and LPG, accounting for 99.3% of its energy demand. This reliance is a significant source of greenhouse gas emissions and deforestation. The proposed study provides a green alternative by utilizing heat that would otherwise be squandered with benefits including avoided cost of energy, through which the use of surplus and wasted heat energy, and reduced investment in individual household or building heating equipment. This paper conducts a feasibility study to assess technical and economic viabilities of introducing a solar water heater reinforced rural heat grid network in the region. The technical feasibility of this system is determined using TRNSYS and Polysun to see if the proposed system provide at least 60% solar fraction in all seasons, while economic analysis is conducted using Benefit Cost Ratio (BCR) and discounted payback period. A pilot project was carried out for five households for demonstration, which showed that the heating demand of each house could be met by 40 different 15-tube solar collectors, resulting in energy savings of 2,28,000 kWh and reducing 25,423 kgs of CO_2 annually. The financial analysis revealed a BCR of 1.45 and a payback period of 14 years, showing that heat grid network is techno-economically viable for rural communities of Nepal.

Keywords. Heat grid, techno-economic feasibility, TRNSYS, Polysun, district heating

1. Introduction

Affordable, feasible, reliable and clean energy sources have a significant impact on economic growth, environment and social well-being. During FY 2019/20 Nepal produced a total of 3021 GWh of electricity registering an increase of 18.57% from FY 2018/19. The 15th Economic Plan of Nepal had the objective to achieve Sustainable Development Goal through sustainable energy usage and promotion of green economy. The strategy was to emphasize the use of latest, sustainable and energy-efficient technologies. By the end of FY 2023/24, it has an aim to achieve 12% ratio of renewable energy in total energy consumption. Similarly, 100% of the population shall have access to electricity, and electricity consumption per capita should have reached to 700 kWh [1]. Residential sector of rural parts of Nepal is predominantly dependent on coal and fuelwood as sources of energy, followed by kerosene and Liquefied Petroleum Gas (LPG) specifically for heating and cooking. Switching to cleaner source of energy with integration of heat grid system can benefit end users through increased energy efficiency, improved environmental condition and attainment of national targets. A rural heat grid, as a heat network, is a concept of centralizing a single power plant to heat water which is piped around different buildings for heating them through radiators and distribute hot water. The use of centralized heating systems allows shift load as per the demand of a certain client and shed load when demand is greater than production. Such systems can be easily maintained, repaired and upgraded. They can utilize any source of energy and combine heat and power for increased efficiency. The status of community heat grid systems varies across communities and countries. The community rural heat grid system model demonstrates how decentralized energy systems can be scaled and adapted to distinct local contexts. Its fulfilment can inspire comparable projects in different regions going through similar challenges, showing that it is feasible to reap sustainable development through the localization [2]. The improved technology leads to a preference for the use of district heating as a way of using renewable energy more efficiently, thereby reducing the use of fossil fuels that contribute significantly to global warming. Additionally, these networks have a lot of potential in terms of decarbonisation. The heat network systems are intricate in nature and it calls for meticulous planning, designing as well as operating that may be problematic to the underprivileged rural dwellers. Consequently, a techno-economic feasibility study should precede before installation and operation. Energy demand and consumption should also be considered for optimal design of the system [2]. The purpose of this research is to carry out a viability study on the technical and economic prospects of establishing a solar water heater integrated rural heating grid in Langtang, Nepal using TRNSYS and Polysun for technical aspects, and Net Present Value (NPV), Internal Rate of Return (IRR), Benefit Cost Ratio (BCR) and Discounted Payback Period for financial analysis.

2. Materials and methods

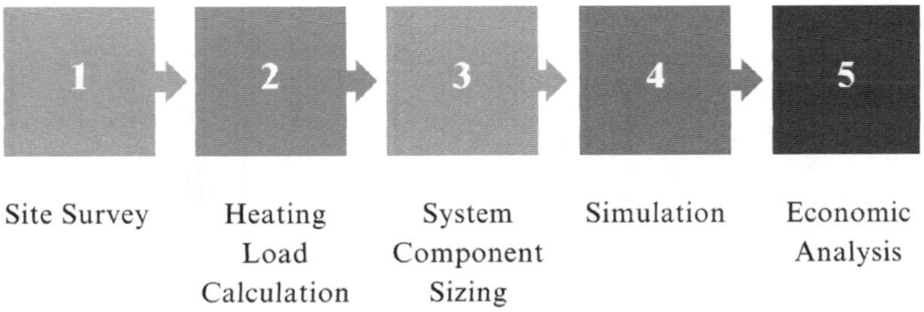

Figure 1. Methodology used in the study.

Figure 1 depicts the methodology used to assess the technical and economic feasibility potential of a solar district heating system to address space heating needs in Nepal's Langtang region. The initial phase involved a comprehensive site survey. Data collection focused on two key aspects: architectural components of potential participating houses and stakeholder engagement. Architectural details, such as building size, insulation levels, and window configurations, were gathered to estimate heating loads.

The survey assessed resident interest and addressed any concerns they might have. Additionally, collaboration with regional coordinators ensured alignment with local development plans and secured their support for the project's implementation. The collected architectural data was then used to estimate the heating load of each house for which a computer simulation using TRNSYS software was conducted.

With the heating load established for each house, the next step involved sizing the various components of the solar district heating system. This involved analytical calculations to determine the capacity of the solar panels, the thermal storage tank, and the distribution piping network. Accurate sizing was essential for ensuring sufficient hot water generation, adequate storage capacity, and efficient distribution throughout the network.

To verify the accuracy of the analytical sizing, a simulation study was conducted using the estimated heating loads and proposed system component sizes. This simulation served as a crucial validation step. By simulating the system's operation over a typical heating season, the research ensured that the proposed design could meet the space heating demands of the participating houses.

Following the successful technical feasibility assessment, an economic analysis was undertaken. This analysis considered the upfront installation costs of the solar district heating system, including the solar panels, storage tank, piping network, and control systems. Additionally, the analysis factored in annual operation and maintenance expenses. To assess the economic viability of the project, potential revenue streams were also explored. These might include cost savings from reduced reliance on conventional heating fuels or potential government subsidies for adopting renewable energy technologies.

2.1 Study site

This study has selected Kyanjin Gompa, a culturally rich village, located at Langtang Valley. It is 9.7 km far from Langtang National Park with 85.5655985° N longitude and 28.2139807° E latitude at an altitude of approximately 3,800 meters above sea level [3]. The village is surrounded by mountains Langtang Lirung (7,227 m) and Yala Peak (5,732 mm). The temperature reaches up to -15 °C in winter season and is high as 11 ^{0}C in the summer as shown in figure 2. It has a total of 40 households with a population of 131 [3]. Its economy is primarily based on tourism, followed by yak herding and agriculture. People usually rely on firewood for heating and cooking purposes, and are mostly engaged in collecting fuelwood from nearby forests causing deforestation in the region.

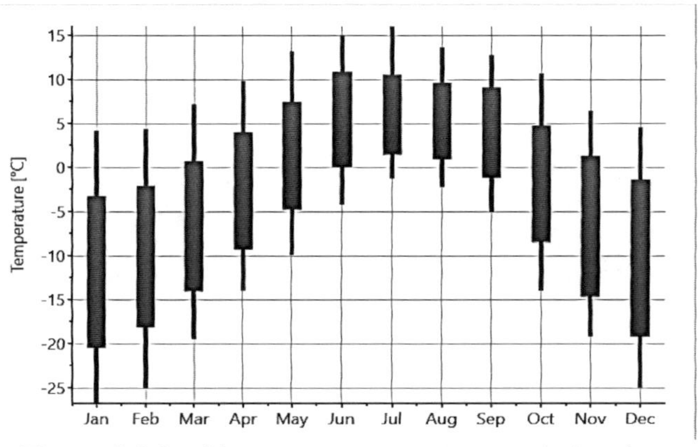

Figure 2. Monthly average temperature variation data.

2.2 Review of existing energy practices in study site

The existing practices in the regions were studied through secondary sources [4,5]. In the Kyanjin Gompa region, the major source of energy in the residential sector are coal and fuelwood, followed by kerosene and LPG for heating and cooking purposes. Shortage of fuel supply and sudden price hikes make it difficult for households and hotels to fulfil the heating demands of residents and tourists. The existing conventional sources of energy such as fuelwood, coal, straw and cow dung are inefficient due to lack of proper design of cookstoves. Modern sources like LPG and kerosene are not affordable to majority of the population. Further, the composition of LPG varies (less propane and more butane), which lowers vapour pressure such that LPG liquid form cannot vaporize.

LPG and kerosene are 50% more expensive than in the capital city, Kathmandu, due to an increase in transportation charges and individual usage. Also, these petroleum products are hugely imported from outside of the country and prices vary according to the international market. Regular supply is challenging due to geopolitical concerns with the neighbouring country, India – the only supplier of petroleum products. Table 1 shows the general overview of existing technology in the region.

Table 1. General overview of existing technology in Kyanjin Region.

Existing Technology	Status	Cons/Problems
Fuelwood	Over 90% of the biomass is predominantly and traditionally used for cooking and heating purposes in households. The main resource of woody biomass is the forest.	• Degradation of Forest • Environmental concerns like Green House Emission
Agriculture Residues	Agricultural residues of crops such as paddy, wheat, millet, jute, etc. are also used as a source of energy particularly for cooking and heating.	• Deterioration of air quality smog, haze, heat waves, and health issues
Commercial Fuels (Petroleum, Coal)	Commercial fuel such as Petroleum products (Kerosene) and coal is almost exclusively consumed by the commercial sector, primarily for heating.	• Limited Resources • Contributes to Environmental Pollution

2.3. Resources available
The parameters to determine potential of solar energy for heating water is found out using Meteonorm Software [6] specifically for Kyanjin Gompa.

2.3.1 Solar potential. Figure 3 shows the diffused radiation and global radiation in Kyanjin Gompa village. The figure below shows that the maximum radiation from the sun hits in May at almost 230 kWh/m^2 and the lowest in February at almost 110 kWh/m^2.

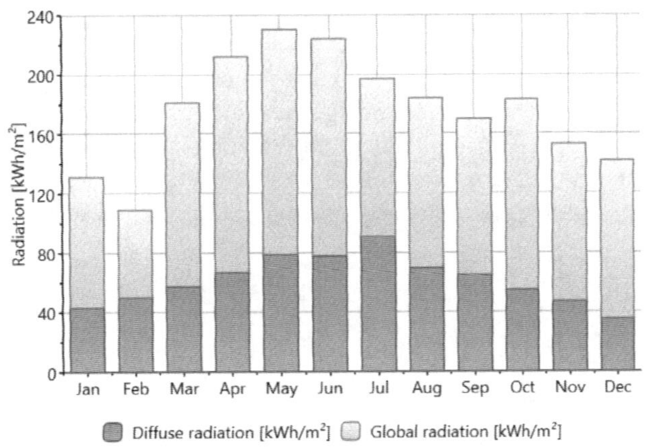

Figure 3. Monthly average solar radiation data.

IOP Conf. Series: Materials Science and Engineering **1314** (2024) 012013 doi:10.1088/1757-899X/1314/1/012013

2.3.2 Sunshine duration. Figure 4 shows that the sunshine duration is usually above 6 hours which makes us see the possibility of the use of the solar mechanism for heating water. As per the figure shown, even at the least there is 7 hours of sunshine in the month of July.

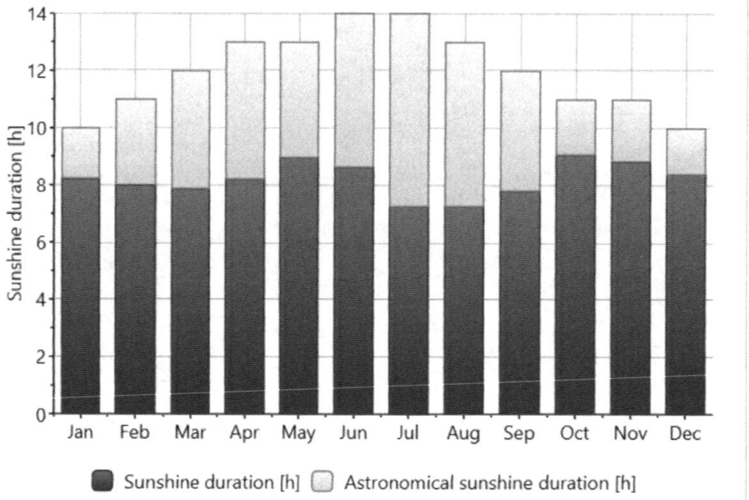

Figure 4. Monthly average sunshine duration variation data.

2.4 System components of rural heat grid system

Figure 5 depicts a schematic of district heating system implemented in the Langtang region of Nepal. The system leverages solar energy to provide space heating for five interconnected houses. The core component is a solar panel that heats water. This heated water is then channelled into a large storage tank with a capacity of 7000 liters. The system continuously circulates water: cold water is pumped into the system, passes through the solar panel for heating, and is subsequently stored in the tank. A controller manages this flow, optimizing circulation and preventing excessive pressure build up. Each house within the network possesses its own 200-liter hot water tank and radiator system. Heated water from the main storage tank is distributed to these individual tanks. The radiators then provide space heating by circulating the hot water.

The system functions as a closed loop. Cooled water returning from the radiators is channelled back to the main storage tank for reheating, ensuring a continuous supply of hot water for space heating. Safety and efficiency are maintained through the incorporation of various control mechanisms, including flow controllers, level switches, and temperature and pressure gauges. These components work in unison to monitor and regulate the system, guaranteeing optimal operation and preventing potential issues.

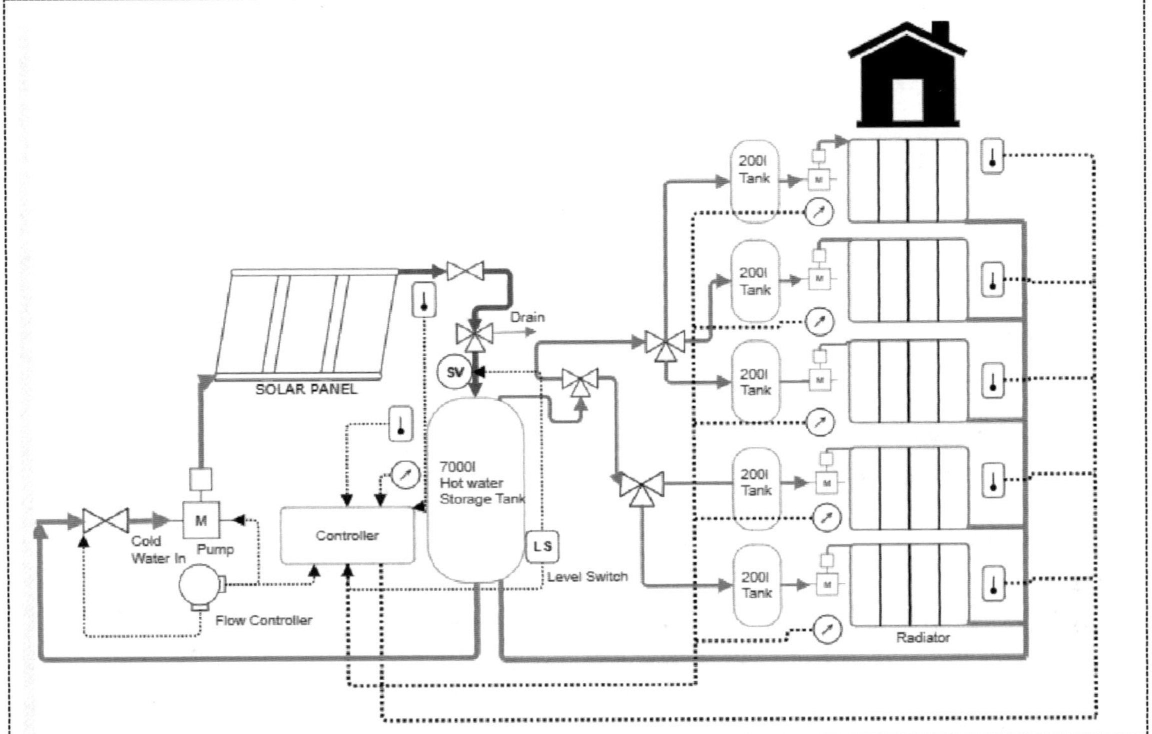

Figure 5. Schematic diagram of rural heat grid.

2.5 System sizing and heat demand

For the system sizing, the heating load of the building for which the system is being designed has to be estimated.

2.5.1 Site survey. A thorough site survey is a prerequisite before the deployment of any Solar Water Heating (SWH) system. The site survey to Kyanjin Gompa primarily focused on identification of heating load of the house throughout the year. So, in order to estimate the heating load, it was necessary to have the information of the architectural components located in the study site. The architectural components included assessment of the roof condition, slope, and type of roof, as these factors influence the selection of the mounting method and the optimal inclination angle for the solar collectors. Comprehending the existing pipe layout is a vital aspect of the survey. This assists in identifying the materials used in the existing piping, their compatibility, and the positioning of hot water tanks, as well as hot and cold-water lines. Moreover, it provides insights into the water head, facilitating efficient integration of the SWH system. The electrical layout is also a significant consideration during the site survey, aiding in pinpointing the best location for the SWH system and the system controller to ensure effective functionality. Besides, immediate environmental factors such as potential shading and the local climate are evaluated during the survey. These factors are important as they can impact the efficiency and performance of the SWH system in our research setting. The site survey includes the evaluation of the hot water demand per house to fulfil the thermal requirements of the individual household

Data collection from the site survey was collected through Key Informant Interview (KII) of Chairperson of Gosaikunda Rural Municipality and the ward secretary and the immediate stakeholders and end users of the system. The overview of the data collection is tabulated in table 2.

IOP Conf. Series: Materials Science and Engineering **1314** (2024) 012013 doi:10.1088/1757-899X/1314/1/012013

Table 2. Data Collection Method

SN	Method	Respondents	No. of Respondents
1	Key Informant Interview	Intermediate Stakeholders and End Users	5 (Representative of 60 households)
2	Semi-Structured Telephone Interview	Regional coordinators and Program Officers	5

During this communication, information regarding the orientations of each room, doors, windows and architectural details of the building, including construction materials, ceiling, type of roof, etc., was obtained. Additionally, details about building occupancy and the activities conducted within the house were documented and floor plan of house was constructed using details provided shown in figure 6 and 7.

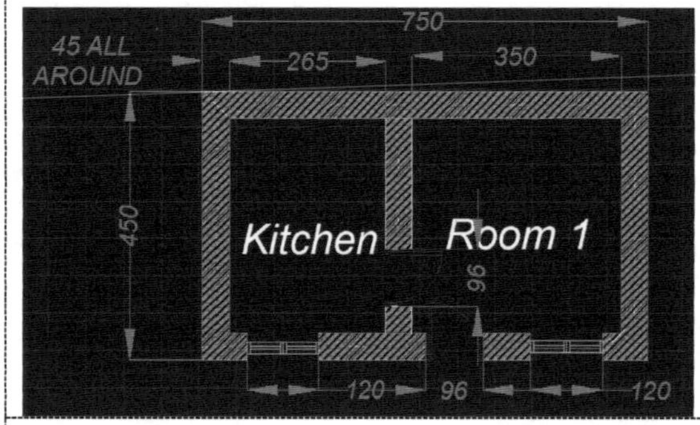

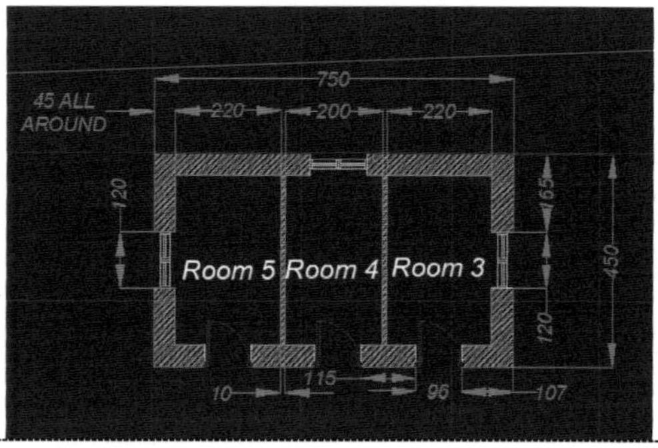

All Dimensions are in cm.

Figure 6. Floor plan of zone 1. **Figure 7.** Floor plan of zone 2.

2.5.2 Building materials. During the site survey, the materials for the different architectural components were obtained as listed in table 4. The U value is calculated from the equation 1.

$$\frac{1}{U} = \frac{1}{hiA} + \frac{L}{KA} + \frac{1}{h_oA} \quad (1)$$

Where, h_o=Convection heat transfer outside the room

A= Surface area of the component

h_i= Convection heat transfer coefficient inside the room

L=thickness of the component

The values of K and L are tabulated in table 3.

Table 3. Thermal Conductivity of material used in a house

S. No	Material	Thickness (m)	K	Reference
1	Stone	0.4	1.33	[7]
2	Clay	0.05	1.26	[7]
3	Plywood	0.1	0.54	[8]
4	Glass	0.05	1	[9]
5	Sal wood	0.2	0.27216	[8]
7	Paint	0.01	0.2	[10]
8	Clay Ground	0.5	1.69	[7]

Table 4. Architectural components with materials.

S. No	Component	Material	Thickness (m)	U-Value (W/m^2K)
1	External Walls	Stone+Clay+Paint	0.45	0.651
2	Internal Wall [Ground Floor]	Stone and Clay	0.45	0.649
3	Internal Wall [First Floor]	Plywood	0.1	1.203
4	Windows	Sal Wood + Glass	0.25	5.6
5	Main Door	Sal wood	0.3	1.203
6	Room Door	Sal Wood + Plywood	0.3	0.665
7	Ceiling	Sal Wood + Clay	0.3	1.104
8	Floor	Clay	0.5	1.316

2.5.3 Heating load. Following the data collection phase, the research process involved creating a detailed computer model of the building to analyze its thermal performance. TRNBuild software [7], a building energy simulation tool, was used to construct a three-dimensional multizone model. This model aimed to replicate the actual building as closely as possible, allowing for virtual experimentation and performance evaluation under various conditions.

The model incorporated the building's orientation relative to cardinal directions, a significant factor influencing solar heat gain. Additionally, the slope of the roof and other architectural components were integrated into the model, as these elements can impact heat transfer through the building envelope. Material properties play a vital role in determining a building's thermal behaviour. The data compiled in table 1 was utilized to define the properties of various building materials, such as walls, roofs, windows, and floors. These material properties encompass characteristics like thermal conductivity, density, and specific heat capacity, which influence heat transfer rates through the building envelope. Within the TRNBuild software, these defined materials were combined to create virtual representations of the building's architectural components.

To further enhance the model's accuracy, the building was divided into distinct thermal zones. This zoning approach acknowledges that different areas within a building can experience varying temperature conditions. The model incorporated two distinct zones: one representing the lower floor and another for the upper floor. This distinction allows for a more nuanced understanding of heat distribution within the building. Within each thermal zone, the model further defined five air node regions. These air nodes represent specific locations within the zone, such as the kitchen, living room, and individual bedrooms (Room 1-5). The dimensions of each air node were assigned based on the corresponding building orientation, as detailed in figures 6 and 7. This level of detail ensures that the model accounts for potential variations in temperature distribution across different rooms due to factors like solar radiation exposure, occupant activity and lighting activity.

A crucial aspect of building energy modelling involves establishing a setpoint temperature, which represents the desired indoor temperature that the heating and cooling systems aim to maintain. The researchers defined a setpoint temperature of 20°C for the entire building. Internal heat gains, which can significantly impact heating requirements, were also incorporated into the model. A cooking schedule was defined and assigned to the kitchen air node (air node 1) to account for the heat generated during food preparation. Similarly, the model considered heat gains from occupants by defining a presence schedule for each air node (Figures 9-13). This schedule reflects the typical occupancy patterns within the building, influencing internal heat generation. Light usage within the building can also contribute to internal heat gains. To account for this factor, light schedules were defined for each air node.

Heat loss through the building envelope is another critical aspect of thermal performance modelling. The model incorporated a constant infiltration value of 0.6 air changes per hour (ach) to account for heat loss due to air leakage through cracks and gaps in the building envelope.

IOP Conf. Series: Materials Science and Engineering **1314** (2024) 012013 doi:10.1088/1757-899X/1314/1/012013

The model was designed to accept various inputs, including weather data for the building site. This weather data encompasses information like solar zenith and azimuth angles, which are crucial for calculating solar heat gain on the building's surfaces. Additionally, the model accepted occupancy profiles for each room, light schedules for each room, and a cooking schedule for the kitchen. These occupancy and usage profiles allow the model to simulate realistic scenarios and predict the building's heating demand under typical operating conditions. The primary output of the model was designed to generate the heating demand for each individual air node within the building. This allows for a detailed understanding of how heat distribution varies across different rooms. Furthermore, the model was equipped with an integrator component that calculated the total heating demand for the entire house. This value provides a crucial metric for assessing the overall heating requirements of the building and informing decisions regarding the design of the solar district heating system.

Figure 8 illustrates the layout of the TRNSYS simulation model, visually depicting the various components and their interactions. The output of the simulation was set as N30 type i.e sensible heating demand of the house and each of the air node.

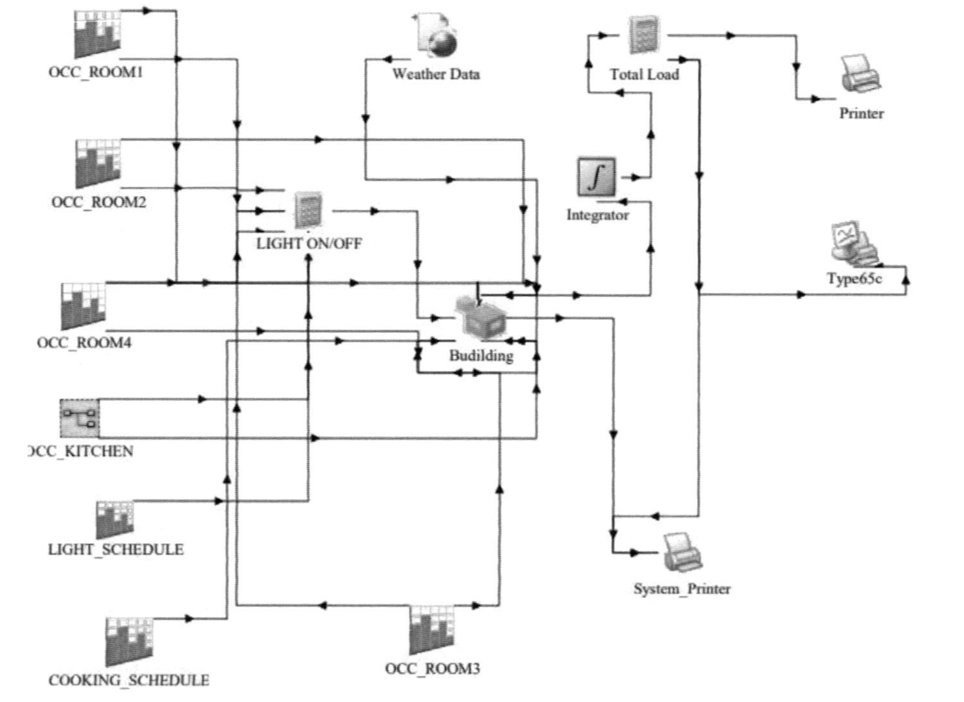

Figure 8. Layout of TRNSYS simulation of a house.

With completion of multizone building process, simulation studio was used to calculate the heating load of a house. The simulation process operated on hourly weather data extracted from Meteonorm in tmy2 format, factoring in both the occupancies within the houses and the heat gain stemming from various electrical equipment.

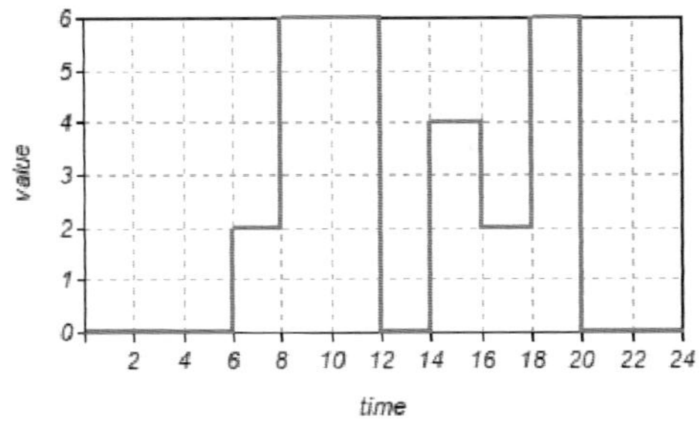

Figure 9. Schedule of people inside kitchen

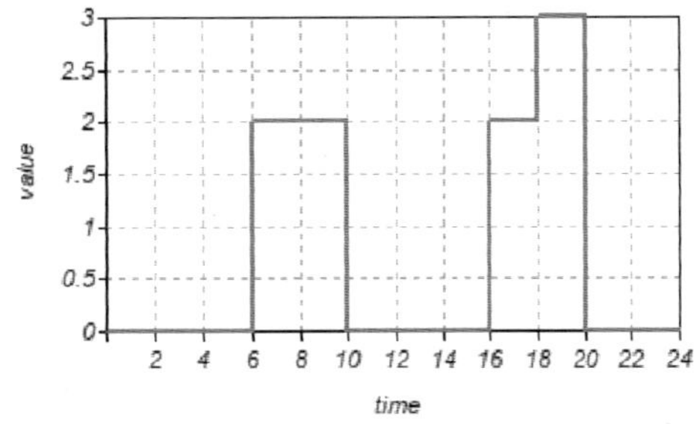

Figure 10. Schedule of lighting inside kitchen

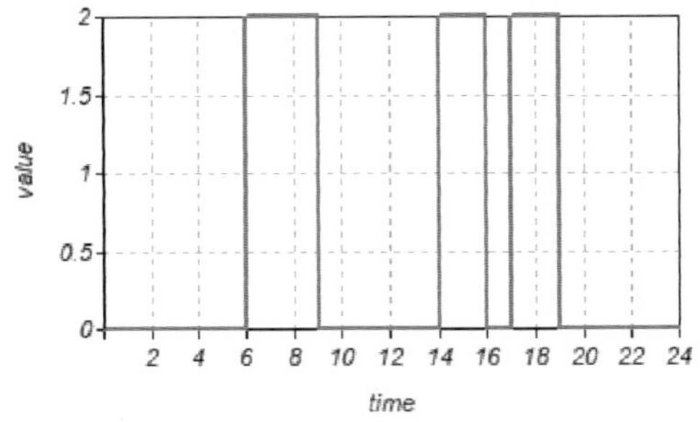

Figure 11. Schedule of cooking in kitchen

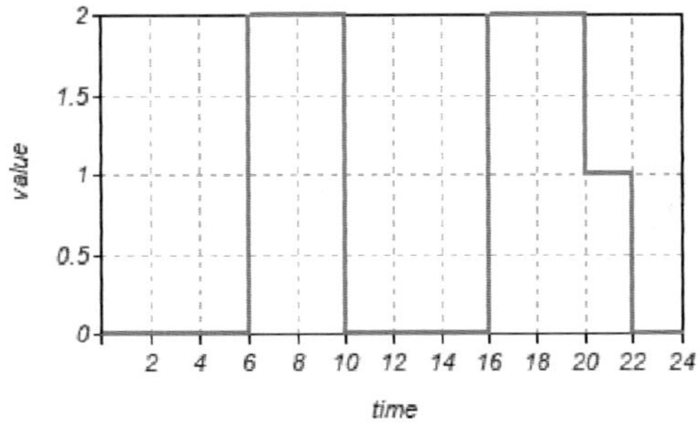

Figure 12. Schedule of lighting inside Room 1-4

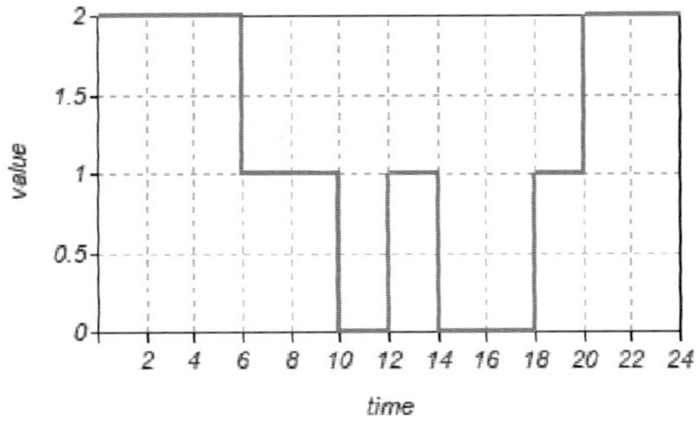

Figure 13. Schedule of people inside Room 1-4

The mathematical equations related to heating load calculation are summarized in table 5.

Table 5. Governing equations to calculate heating load

Component	Heat Gain	Reference
External Wall, Roof, Window	$Q = UAdT$	[11]

IOP Conf. Series: Materials Science and Engineering **1314** (2024) 012013 doi:10.1088/1757-899X/1314/1/012013

Partitions (Internal Wall)	$Q = UAdT$	[11]
People Heat Load	$Q_s = \text{Sensible}\frac{\text{heat}}{\text{people}} \times \text{No of peope}$	[11]
	$Q_L = \text{Latent}\frac{\text{heat}}{\text{people}} \times \text{No of peope}$	[11]
Lighting Heat Load	$Q = \frac{\text{watts}}{m^2} \times \text{area}$	[11]
Infiltration	$Ql = 0.68 * CFM * dW$	[11]
	Where, CFM=Crack Length*0.37 (For Window) CFM=Crack Length*0.5 (For residential door) dW=outdoor humidity ratio	

3. System component sizing

The determination of component sizes for the system is intricately tied to the heating load, forming the foundation of the sizing process. The calculations and specifications for each component are derived from a thorough assessment of this heating load. In essence, the heating load acts as a fundamental parameter that guides the precise sizing and selection of components within the system. By aligning the capacities and specifications of components with the heating load, the system can deliver the necessary thermal output to meet the heating demands of the building or houses, guaranteeing an optimal and functional heating solution.

3.1 Evacuated solar tube collectors

To know the number of solar collectors needed the useful gain of a 2 m^2 collector was calculated under the weather condition of site. Simulation studio was used to model the collector performance under hourly weather data of Kyanjin Gompa. Figure 14 shows layout for useful energy gain of collector.

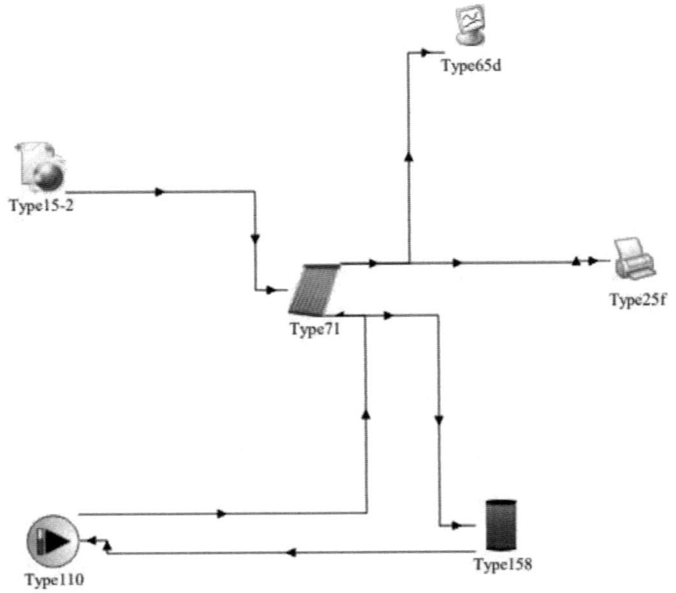

Figure 14. Layout for useful energy gain of collector.

3.2 Pumps

To size the pump, mass flow rate of water and head are required. This can be calculated with the reference of heating load of the room using equation 2.

$$Heating\ load = \left(\frac{m}{t}\right) \times s \times dT \quad (2)$$

3.3 Tank

The capacity of the tank is calculated on basis of the demand for hot water per day using equation 3 [12].

$$capacity\ of\ tank = Water\ demand\ per\ day \times 1.25(Safety\ Factor) \quad (3)$$

3.4 Radiator

Standard 1000 W radiator also known as K-22 type of radiator was taken as reference. Size of the radiator depends upon heating load of a room and is represented by equation 4.

$$Size\ of\ radiator = \frac{Heating\ load\ of\ air\ node}{1000} \quad (4)$$

3.5 Calculation

Heating load of a single house shown in figure 15 was obtained to be maximum of 5.15×10^6 KJ/hr which corresponds to 1440 KW of thermal power needed to heat a single house in a month. For 12-hour duration, the average heating load is around 60 KWh. For piloting of five houses, design heating load of the system is 300 KWh.

IOP Conf. Series: Materials Science and Engineering **1314** (2024) 012013 doi:10.1088/1757-899X/1314/1/012013

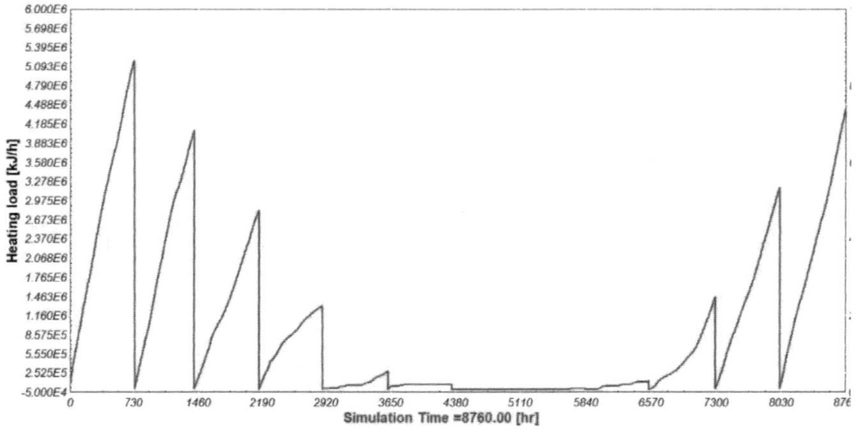

Figure 15. Heating load of a single house.

Useful gain of the system was obtained to be 2200 KJ/hr as shown in figure 16 which corresponds to 0.61 KW thermal power generated by 2 m^2 collector area. To fulfil demand of 60% of demand in worst period of year i.e. January, 40 evacuated solar thermal collectors were calculated which provided collector area of 100 m^2 generating 200 KWh energy when heated for 8 hrs of sunlight with solar irradiance of 4.5 $kWh/m^2/day$.

$$Mass\ flow\ rate\ of\ system = \frac{28 \times 10^3}{4200 \times 50}$$

Pump with the flow rate of 8 l/min was calculated for total hot water demand (12 hr heating) of 5600 l. So, a tank of 7000 l capacity should be installed, based on market available size. With the average heating load of a room averaging around 1 KW, one radiator is enough for the room.

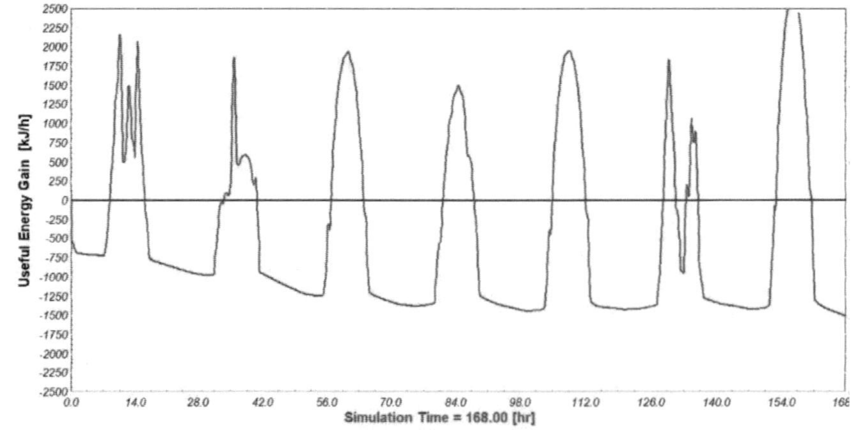

Figure 16. Useful energy gain of a collector.

3.6 Simulation

Following the system component sizing, the research team integrated the resulting data into Polysun simulation software [8] via a trial license obtained from Vela Solaris. The software allows for modelling and performance evaluation of solar energy systems.

The first step involved specifying the system location. The model incorporated the geographic coordinates of Kyanjin Gompa i.e 85.5655985° N longitude and 28.2139807° E latitude.

Next, a system model was constructed within Polysun using its built-in features and tools. The software facilitated the adjustment of various components to precisely match the data obtained from the earlier calculations.

Following a review and verification of all settings and configurations, the simulation was initiated. The simulation regime within Polysun encompassed several variants, input conditions, and output conditions for the rural district heating system in Kyanjin Gompa.

- Scenarios: The simulation considered multiple scenarios. The base case scenario represented the existing situation with no solar heating system, relying solely on wood for space heating. The proposed system scenario incorporated solar thermal collectors and a district heating network, aiming to reduce wood consumption and associated costs. Additionally, alternative configurations were explored, involving variations in the number and types of solar collectors, as well as different storage tank sizes.
- Input Conditions: The software received various location-specific data for Kyanjin Gompa, including longitude, latitude, elevation, and weather details such as average outdoor temperature, global irradiation, and diffuse irradiation. System component details were also provided. These included:
  - ➢ Solar collectors: 40 collectors with a total gross area of 83.2 m², a tilt angle of 40°, and a south-facing orientation.
  - ➢ Pumps for different loops, each with specific flow rates and electricity consumption.
  - ➢ Buffer tank: volume of 7,000 liters.
  - ➢ Energy demand: annual heating energy requirement of 76715 kWh, with a setpoint room temperature of 20°C.
- Output Conditions: The simulation results provided a comprehensive overview of energy production and consumption. Key outputs included:
  - ➢ Solar field yield: 466622 kWh annually.
  - ➢ Solar fraction: percentage of energy provided by the solar collectors.

4. Economic analysis

For economic analysis, a dual approach was adopted: drawing upon simulation software Polysun and leveraging theoretical insights from the textbook by Sullivan. Initially, the estimation of the total investment required for the proposed model involved consulting prevailing market rates and online sources. The assessed total investment for establishing the rural heat grid for five houses stands at NRs. 4,000,000. The breakdown is shown in figure 17. To ensure a comprehensive economic outlook, the operating and maintenance costs of the system were calculated at NRs. 500,000 annually. Figure 18 shows cash flow for 20 years. These costs encompass various components, including staff salaries, depreciation, electricity and miscellaneous expenses.

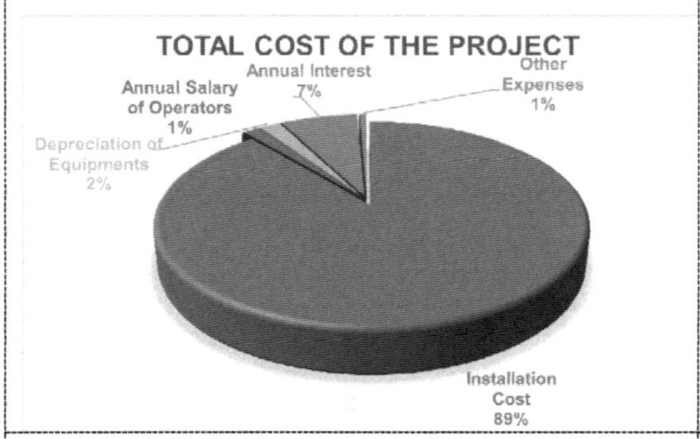

Figure 17. Total cost of the project.

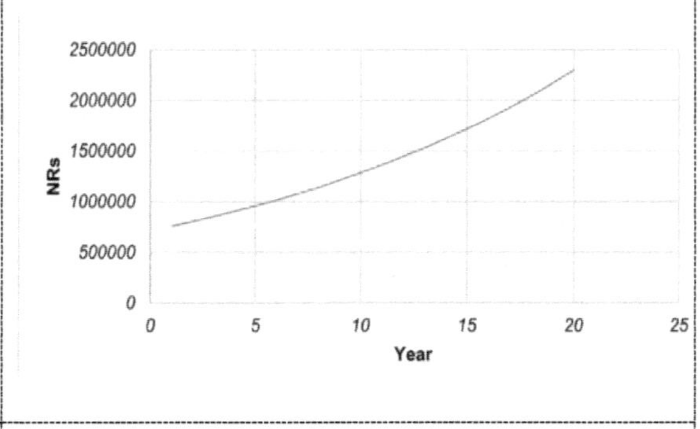

Figure 18. Cash flows over the years.

To quantify the potential economic benefits, a thorough analysis of existing practices was undertaken, focusing on wood consumption for space heating purposes. Drawing from real-world examples, it was established that a commercial home typically utilizes 200 kg of wood per bed per year, amounting to approximately 1.5 kg of wood per day during operation (averaging around 200 days) [5]. Extrapolating from these figures, a commercial establishment with eight beds burns around 2400 kg of wood annually. From a technical standpoint, 1 kg of wood was observed to produce approximately 8.2 MJ [13] of energy, equivalent to 2.27 kWh. The commercial establishment had been spending NRs 1200 for 40 kg of wood, implying that the cost for 1 kWh of energy was NRs 13.33. To meet the heating load, an expenditure of around NRs 10,00,000 was required, given that the annual heating load of the establishment was approximately 76000 kWh. The Rural Heat grid, a community-owned system, charged half of the electricity price, i.e., 2 NRs/kWh. This amounted to a cash flow of NRs 760,000, which will experience a 6% inflation increase per year.

4.1 Initial Cost Burden and Financing Options

The initial establishment cost of the rural heat grid system in Kyanjin Gompa, estimated at NRs. 4,000,000, presents a significant financial hurdle for the community. This substantial upfront investment, particularly in the context of limited funding options in rural Nepal, could hinder project adoption.

To address this challenge, various funding sources are explored whose objective is to promote renewable technology in rural Nepal. Potential programs are listed in table 6.

Table 6. Potential Funding Agencies

S. N	Name of Organization	Targeted Areas	Objective	References
1	Alternative Energy Promotion Centre (AEPC) / UNDP (RERL)	Rural Nepal (micro hydropower projects, solar mini grids, solar pumps)	Remove barriers to renewable energy utilization, support economic, environmental, and social development, reduce greenhouse gas emissions	[14]

2	UNDP	Rural electrification, women's empowerment	Promote renewable energy, support rural electrification, empower women	[15]
3	Global Funds (Green Climate Fund, GEF, Adaptation Fund, Least Developed Countries Fund, NAMA Facility)	Rural and underdeveloped areas	Support sustainable development, climate resilience, and renewable energy projects	[16]
4	Asian Development Bank (ADB)	Rural electrification, renewable energy projects	Fund renewable energy projects, support rural development	[14]
5	World Bank	Renewable energy, sustainable development	Prioritize investment in green, resilient, and inclusive development	[17]
6	European Union	Renewable energy projects, rural development	Support renewable energy initiatives, promote sustainable development	[17]
7	United Nations	Various development projects	Support sustainable and inclusive development, promote renewable energy	[16]
8	Embassies of Various Countries	Renewable energy and development projects	Provide financial and technical assistance for renewable energy projects and development	[16]

A detailed calculation of the payback period is depicted in figure for project evaluation. This calculation illustrates the timeframe required for cost savings, achieved through reduced wood consumption and lower heating expenses, to offset the initial investment. Figure 19 shows that on an ideal situation, the payback period can be achieved during the 14th year which can be even faster if soft loans or funding can be achieved during the construction of the project.

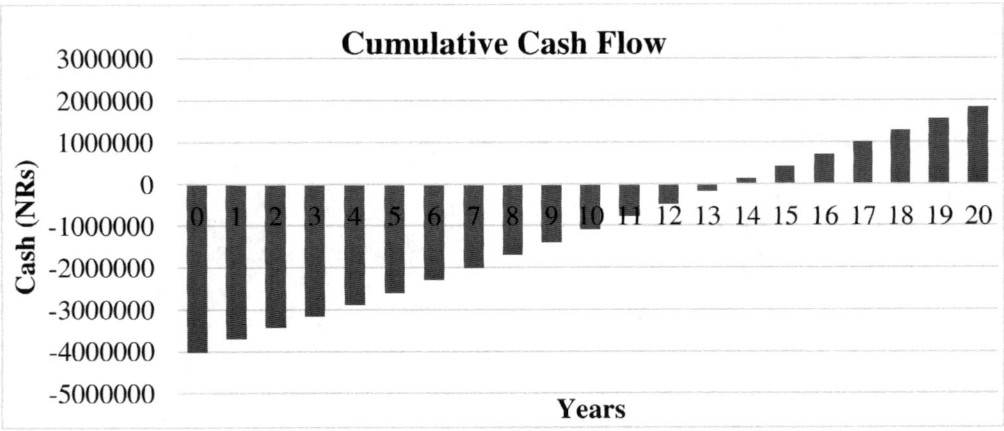

Figure 19. Payback period of the project

IOP Conf. Series: Materials Science and Engineering **1314** (2024) 012013 doi:10.1088/1757-899X/1314/1/012013

4.2 Maintenance and Operations

The annual operating and maintenance cost for the rural heat grid system is estimated at NRs. 500,000. Figure 20 depicts that the components incorporated such as annual salary of operators, depreciation of equipment, repair and maintenance, annual interest and other cost. The annual interest covers major part followed by depreciation of equipment and annual salary of operators. The successful operation of the rural heat grid system requires skilled personnel. It's crucial to address whether local staff can be trained to manage and maintain the system or if external expertise will be needed. Training local staff can enhance sustainability and ensure quick response to any issues that arise. For the potential risk of the downtime backup systems, regular maintenance schedule and spare parts inventory can be maintained.

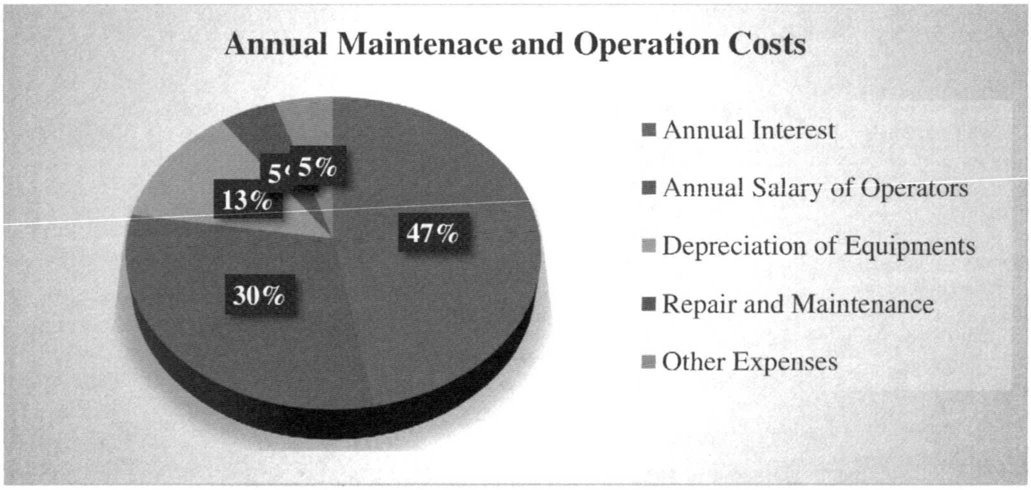

Figure 20. Annual Maintenance and Operation Costs

5. Results

5.1 Heating load of house and each air node

From the simulation studio, it was obtained that heating load was maximum for kitchen and Room 4 are greater compared to other air nodes as shown in figure 21-26 and listed in table 7. Notable points from these diagrams are that heating loads are for the half of the year which lies in between October to March. Figure 18 shows that total heat capacity of the system is highest during January and gets minimum during months June to August with minimal heat requirements.

Table 7. Heating loads of each air node

S. No	Regime	Purpose	Heating Load
1	Kitchen	Cooking	Maximum Load: 2.2 KW
			Minimum Load: 0.33 KW
2	Room 1	Dining	Maximum Load: 0.99 KW
			Minimum Load: 0.06 KW
3	Room 2	Bedroom	Maximum Load:1.01 KW
			Minimum Load: 0.03 KW
4	Room 3	Bedroom	Maximum Load:0.97 KW
			Minimum Load:0.04 KW
5	Room 4	Bedroom	Maximum Load: 1 KW
			Minimum Load: 0.03 KW

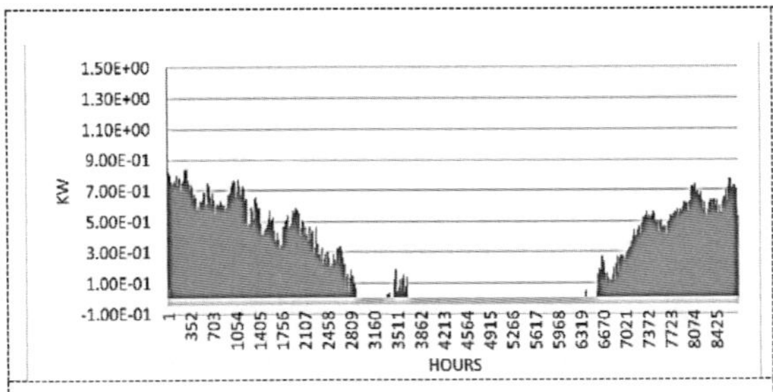

Figure 21. Heat load of kitchen.

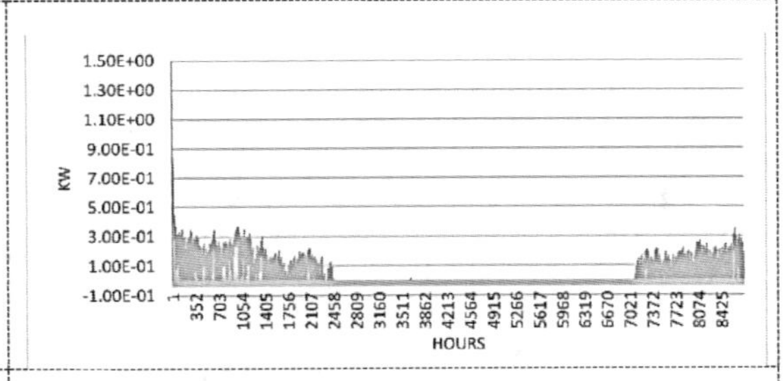

Figure 22. Heat load of Room 1.

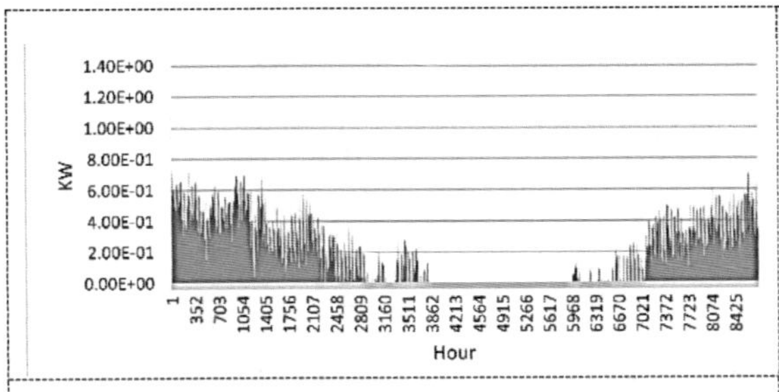

Figure 23. Heat load of Room 2.

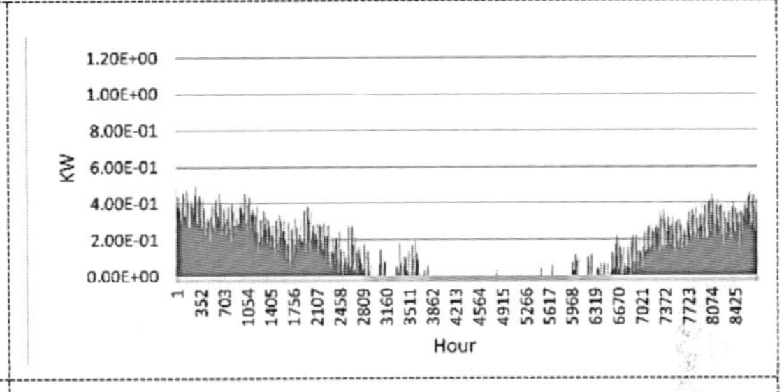

Figure 24. Heat load of Room 3.

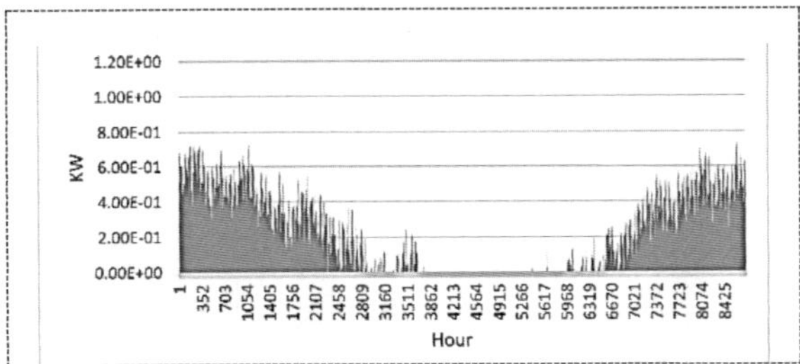

Figure 25. Heat load of Room 4.

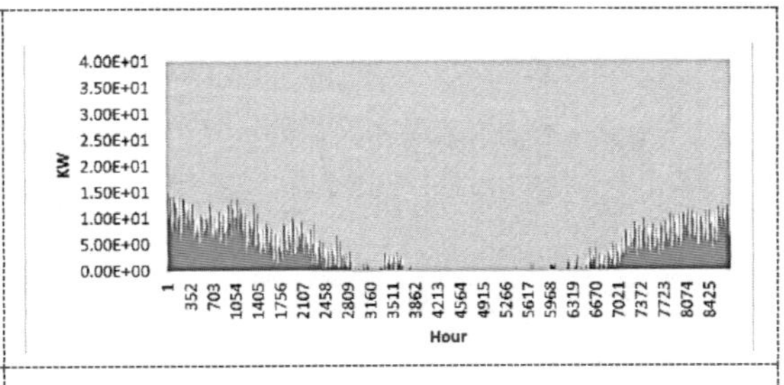

Figure 26. Heating load of the system per hour.

5.2 Simulation result

Figure 27 and 28 depicts the total thermal energy provided by collector to the system and has shown that average of 60% load is fulfilled by solar thermal energy.

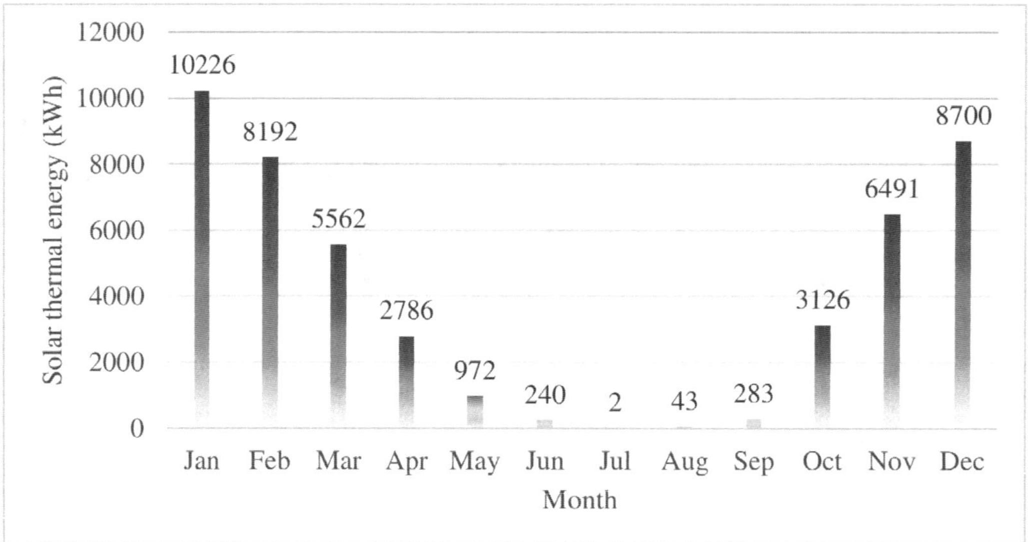

Figure 27. Solar thermal energy generated by collector per month.

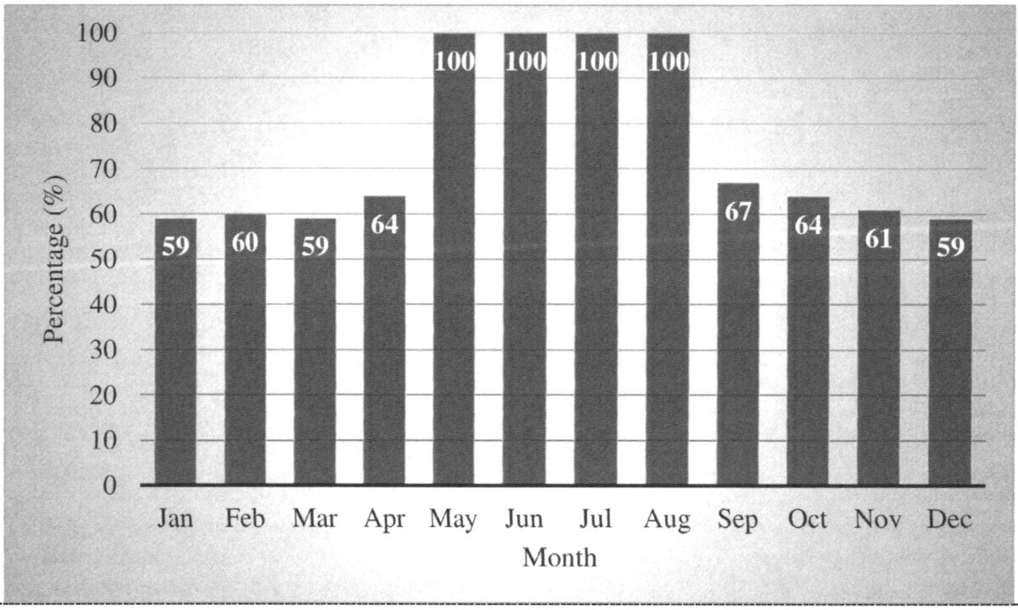

Figure 28. Solar fraction of the system per month.

5.3 Economic viability

The economic assessment was conducted over a timeline of 20 years, which was the anticipated duration of the project. A detailed analysis was performed, and the net present value (NPV) at the end of this period was calculated to be NRs. 18,54,897. This calculation acted as a crucial indicator, confirming the project's feasibility and potential for investment. The system's internal rate of return (IRR) was determined to be 14.19%, which exceeded the prevailing interest rate, indicating the project's economic viability. This was further supported by a Benefit/Cost ratio of 1.45 and a payback period of 14 years.

IOP Conf. Series: Materials Science and Engineering **1314** (2024) 012013 doi:10.1088/1757-899X/1314/1/012013

6. Discussion

The Community Rural Heat Grid System stands as a transformative energy solution meticulously crafted to address diverse challenges encountered by rural communities. This innovative system serves as a clean alternative, supplanting traditional heating methods and effectively mitigating indoor air pollution. By doing so, it plays a crucial role in alleviating respiratory ailments and other health issues linked poor indoor air quality.

Notably, the implementation of this system also yields a notable enhancement in overall safety within the community. It achieves this by significantly reducing the incidence of heating-related accidents, such as fires, burns, and carbon monoxide poisoning. It provides a clean alternative the traditional methods like burning coal, wood, which are the major source of greenhouse emission in the region. This transition from conventional methods not only ensures a safer living environment but also results in more sustainable fuel consumption and cost reduction, rendering the system an efficient and economically viable heating solution. The system also addresses the fundamental health and safety issues by mitigating the indoor air pollution and reducing fire hazard associated with traditional heating methods.

A key aspect of this transformation lies in the substantial annual energy savings achieved through the utilization of solar thermal energy. The community rural heat grid makes an appreciable contribution in reducing greenhouse gas emission and promoting environmental sustainability. The system results in annual savings calculated to be a maximum of 228,000 kWh. Such significant energy savings have a direct positive impact on the environment by considerably reducing greenhouse gas emissions and raising environmental profile. The system effectively curtails the community's carbon footprint, resulting in a substantial reduction of up to 25,423 kg of CO_2 emissions per year. These figures align a clear indicator of the system's positive environmental impact with global sustainability goals that underscores the system's effectiveness in decreasing the community's carbon footprint, promoting both environmental and public health benefits for rural communities.

The Community Rural Heat Grid System is an excellent instance of a model for a sustainable development that yields significant advantages for the environment, economy, and community. The Community Rural Heat Grid System's provides a model for scaling and reproducibility in other areas. Its genesis and implementation offer useful insights in renewable energy systems and offers useful perspectives on the difficulties and remedies related to renewable energy systems in other rural regions. The Community Rural Heat Grid System provides a blueprint for future projects with similar and comparable social, economic, and environmental advantages. In addition, the Community Rural Heat Grid System creates opportunities for more study and advancement. Its accomplishments demonstrate potential for expanded horizons in current technology as well as exploration of new ways to address various local scenarios. The system acts as a case study, offering useful information and insights that can guide future advancements in sustainable development and renewable energy. The system makes a strong case for community support due to its effects on lowering greenhouse gas emissions, enhancing public health, and producing financial savings. Furthermore, the system's proven viability might draw funding from the public and commercial sectors, encouraging the creation and implementation of other initiatives along these lines.

7. Conclusion and recommendation

Community Rural Heat Grid System emerges as a holistic and transformative solution tailored to address the multifaceted challenges faced by rural communities, particularly in regions like Kyanjin Gompa, Nepal. The innovative system not only provides a clean alternative to traditional heating methods but also serves as a catalyst for improved health, safety, and economic well-being.

One of the system's key strengths lies in its role as a sustainable energy solution, leveraging solar thermal energy to meet the heating demands of households. The simulation results showcase its efficiency, with an average of 60% of the heating load being fulfilled by solar thermal energy. This not only reduces the reliance on traditional energy sources but also contributes significantly to environmental conservation by mitigating greenhouse gas emissions. The projected reduction of up to

25,423 kg of CO_2 emissions per year aligns with global sustainability goals, making the system a commendable step towards a cleaner and greener future. Beyond environmental benefits, the Rural Heat Grid System addresses critical health and safety concerns prevalent in rural areas. By replacing conventional heating methods, such as wood and coal, the system mitigates indoor air pollution, alleviating respiratory ailments and promoting better overall health. Additionally, the reduction in fire hazards associated with traditional heating methods ensures a safer living environment for the community.

Economically, the system demonstrates viability and sustainability. The economic analysis, spanning a 20-year timeline, reveals a positive net present value (NPV) and an internal rate of return (IRR) exceeding the prevailing interest rate. The Benefit/Cost ratio of 1.45 and a payback period of 14 years further affirm the economic feasibility of the project. This indicates that the investment in the Rural Heat Grid System can yield substantial returns over its operational lifespan, making it an attractive proposition for investors and stakeholders.

The successful implementation of the system hinges on its adaptability to the local context, considering factors such as building structures, heating loads, and economic conditions. The site survey conducted in Kyanjin Gompa provided crucial insights into the specific needs and preferences of the community, facilitating the customization of the system to suit their requirements. In essence, the Community Rural Heat Grid System exemplifies a sustainable and community-driven approach to addressing energy challenges in rural Nepal. It integrates technological innovation with environmental stewardship, health improvement, and economic viability, setting a precedent for similar initiatives in other regions grappling with similar issues. As rural areas continue to seek solutions that balance progress with environmental responsibility, the Rural Heat Grid System stands as a beacon of sustainable development and a model for inclusive, community-centric energy solutions.

Acknowledgment

The authors are grateful to Efficiency for Access, Energize-Nepal Program, Residence of Kyanjin Gompa and Project 22-05 Laboratory, Kathmandu University.

References

[1] NPC 2019 The Fifteenth Plan (2076/77-2080-81) National Planning Commission, Government of Nepal Kathmandu Nepal 1–418
[2] Shah M, Koirala K, Upreti S and Sanjel N 2023 Techno-Economic Analysis of Energy Systems in Thakle Namuna Basti - A Case Study *IOP Conf Ser Mater Sci Eng* **1279** 012007
[3] Full Time Explorer Kyanjin Gompa, Nepal: Village Guide for Trekkers ⋆ Full Time Explorer
[4] Lama N B 2019 *ECONOMIC DIMENSION OF TOURISM IN LANGTANG VILLAGE Central Department of Rural Development in partial fulfillment of the requirements for the Degree of Master of Arts in Rural Development*
[5] Chapagain P S 2017 Firewood management practice by hoteliers and non-hoteliers in Langtang valley, Nepal Himalayas *Geographical Journal of Nepal* **10** 55–72
[6] Meteonorm intro - Meteonorm (en)
[7] Transient System Simulation Tool 2018 *TRNSYS 18*
[8] Amesweb Thermal conductivity of Hardwoods and Softwoods
[9] Hyde T J, Arya F and Hewitt N 2014 *A novel building component hybrid vacuum glazing-A modelling and experimental validation*
[10] OTM Solutions Thermal conductivity of thin materials (paint, coating, metal sheet) – OTM Solutions Pte Ltd
[11] Spitler J D and ASHRAE (Firm) *Load calculation applications manual*
[12] Goverment of Kenya *Solar Water Heating Training Manual for the Kenyan Industry*
[13] Food and Agriculture Organization of the United States Sustainable Forest management
[14] Renewable Energy for Rural Livelihood Renewable Energy for Rural Livelihood (RERL) | GEF

[15] United Nations Development Programme Funding & Delivery | United Nations Development Programme
[16] Ministry of Energy W R and I SDG7 Energy Compact of Ministry of Energy, Water Resources and Irrigation, Government of Nepal A next Decade Action Agenda to advance SDG7 on sustainable energy for all, in line with the goals of the Paris Agreement on Climate Change
[17] World Bank Nepal: Government, Development Partners Prioritize Investment in Green, Resilient, and Inclusive Development

AUTHOR INDEX